T0214035

High Performance Simulation for Industrial Paint
Shop Applications

Kevin Verma • Robert Wille

# High Performance Simulation for Industrial Paint Shop Applications

 Springer

Kevin Verma
Johannes Kepler University
Linz, Austria

Robert Wille
Johannes Kepler University
Linz, Austria

ISBN 978-3-030-71627-1     ISBN 978-3-030-71625-7   (eBook)
https://doi.org/10.1007/978-3-030-71625-7

This Springer imprint is published by the registered company Springer Nature Switzerland AG
The registered company address is: Gewerbestrasse 11, 6330 Cham, Switzerland

# Preface

In the 1960s, scientists initially started to use computer algorithms to simulate rudimentary fluid flow scenarios. Nowadays, the study of fluid flows is recognized as an essential part of *Computer Aided Engineering* (CAE). Corresponding tools are based on the methods for *Computational Fluid Dynamics* (CFD), which essentially employ numerical analysis to solve fluid flow problems. These tools allow to simulate scenarios that are difficult or costly to measure in the real world, which is why they are extensively utilized in both science and industry. One important industrial application of CFD is the automotive paint shop, which is one of the key processes in automotive manufacturing. It includes various applications for cleansing, corrosion protection, or coating of car bodies.

However, despite these benefits, the simulation results are frequently still not as reliable as physical measurements. This is often caused by the high numerical complexity of CFD methods, where the accuracy of the solution is directly related to the available computer resources. Especially for the simulation of real world scenarios, there is an inherent trade-off between accuracy and computational time.

This book aims to improve on these shortcomings, to eventually render corresponding CFD methods even more practically relevant and applicable to real world phenomena. To that end, the main emphasis is on optimizing these methods by employing *High Performance Computing* (HPC) techniques. Different CFD methods are investigated from an HPC perspective, which results in a variety of methodologies and approaches that allow to conduct corresponding simulations in a significantly more efficient manner. More precisely, this book provides optimized solutions for different fundamental CFD methods, which includes:

- grid-based methods and
- particle-based methods

as well as an alternative method that aims to improve certain aspects of these well-established approaches, namely:

- volumetric decomposition methods.

The efficiency and scalability of the proposed methods are confirmed by both academic and industrial use cases. These industrial scenarios include a variety of applications from the automotive paint shop. It is shown that the proposed methods allow to simulate corresponding applications in significantly reduced computational time. Moreover, the proposed methods allow to consider more complex paint shop applications, which could not be considered by methods available thus far. By this, this book significantly improves on the current state of the art and further contributes to the envisioned goal of making CFD methods more efficient and applicable to real world scenarios.

Overall, this book is the result of several years of research conducted at the *Johannes Kepler University Linz* (JKU), Austria, and the *ESS Engineering Software Steyr GmbH* (ESS), Austria. For always providing a stimulating and enjoyable environment, we would like to thank all members of the Institute for Integrated Circuits and the LIT Secure and Correct Systems Lab at JKU as well as the entire team at ESS. We are particularly indebted to Martin Schifko who strongly supported us in the developments that provided the basis for this book! Furthermore, we sincerely thank the coauthors of all the papers that formed the basis of this book. For funding, we thank the Austrian Research Promotion Agency (FFG) that supported us within the scheme "Industrienahe Dissertationen" under grant no. 860194, the State of Upper Austria that supported us through the LIT Secure and Correct Systems Lab, as well as by the BMK, BMDW, and, again, the State of Upper Austria for support within the frame of the COMET program (managed by the FFG). Finally, we would like to thank Springer Nature and especially Charles "Chuck" Glaser for publishing this work.

Linz, Austria                                                         Kevin Verma

Linz, Austria                                                         Robert Wille
February 2021

# Contents

# Part I
# Introduction and Background

# Chapter 1
# Introduction

Fluids, especially air and water, have a major impact in our daily life. *Fluid mechanics* is the scientific discipline which studies the mechanics of fluids, i.e. how they behave both, at rest and in motion when forces act on them [60]. The motion of fluids (i.e. *fluid flows*) plays a key role in a wide range of applications in both, science and industry. Such fluid flows are usually difficult to measure in the real world, which is why the simulation (i.e. prediction) of fluid flows has increasingly become an important task in engineering and science. These simulations allow to predict complex fluid flow scenarios and, by that, assess various properties of the fluid, such as flow velocity, density, pressure or temperature. Corresponding solutions find applications in a wide range of areas ranging from the medical sector, where they are e.g. applied to simulate arterial characteristics [105], to industrial engineering.

An important industrial application in which fluid flows play a crucial role is the automotive paint shop, which is one of the key processes in automotive manufacturing. It includes various processes for cleansing, corrosion protection, or coating of car bodies. In all of these processes, fluid flows play a key role, e.g. when car bodies are dipped through large coating tanks filled with liquid, or robots spray liquid by high pressure nozzles onto the bodies for cleansing purposes. Since these processes involve substantial material and energy costs, there is a high demand to simulate and, by this, optimize them prior to their implementation.

While nowadays such fluid flow simulations are recognized as essential tools in corresponding disciplines, the study of fluid flow problems dates back to the early 1930s, where engineers were already using basic fluid mechanics equations to solve rudimentary fluid flow problems. However, due to the lack of computing power, they were solved by hand and, hence, significantly simplified. During these times, the study of fluid flows was largely considered theoretical and exploratory [2]. Later in the 1960s, when the computing technology improved, corresponding equations were increasingly solved by computer algorithms [2]. Since then, numerous research emphasized on mathematical models and computer programs with the goal to make

K. Verma, R. Wille, *High Performance Simulation for Industrial Paint Shop Applications*, https://doi.org/10.1007/978-3-030-71625-7_1

the numerical study of fluid flows practically relevant. Nowadays, the study of fluid flows is recognized as an essential part of *Computer Aided Engineering* (CAE) and corresponding tools are extensively utilized in both, science and industry.

Many of these solutions are based on methods for so-called *Computational Fluid Dynamics* (CFD [2]). The core of these CFD methods relies on the *Navier-Stokes-Equations*, which essentially describe the motion of fluids. These equations have been known for 150 years and are named after the French mathematician Claude-Louis Navier and the Irish physicist George Gabriel Stokes [16]. These widely-used simulation tools are based on different dedicated CFD methods, which can be fundamentally categorized as follows:

- **Grid-based methods**: Grid-based methods are the most common and widely-used CFD approaches. Here, the fluid is composed of fluid cells, aligned in a regular grid, each of which contains some given volume of fluid (as visualized in Fig. 1.1a). The grid is fixed in space and usually does not move nor deform with time. The material or fluid moves across these fixed grid cells, while the fluid itself is constrained to stay with the grid. Popular grid-based methods include *Finite Volume Methods* (FVM, [98]) or *Finite Element Methods* (FEM, [82]).
- **Particle-based methods**: In particle-based methods, the fluid is discretized by sample points, which are identified as particles (as visualized in Fig. 1.1b). These particles completely define the fluid, which means that the particles move with the fluid. Some of the most common methods belonging to the particle-based method family are *Moving Particle Semi-implicit* (MPS, [103]), *Finite Volume Particle Method* (FVPM, [37]), or *Smoothed Particle Hydrodynamics* (SPH, [28]). All of them are relatively new and all these methods are closely related.

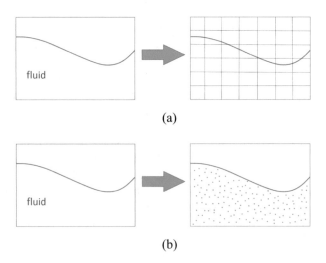

**Fig. 1.1** Basic approaches for fluid flow modeling. (**a**) Grid-based fluid structure. (**b**) Particle-based fluid structure

In addition to these main approaches, there exist also alternative methods which try to improve on certain aspects on these well-established approaches. One of these are so-called *volumetric decomposition methods*:

- **Volumetric decomposition methods**: The main idea of volumetric decomposition methods is to use fewer and larger volume units compared to standard grid-based CFD methods in order to reduce the computational complexity. As such, these methods can be classified as grid-based methods, even though the employed volumetric representation differs to classical methods. In volumetric decomposition methods, a geometrical decomposition into so-called *flow volumes* is applied. As a result, a graph is constructed which represents the topology of the object by flow volumes with their respective relations. The resulting graphs can be seen as reeb graphs [21], which are originally a concept of Morse theory [56], where they are used to gather topological information.

Each of these approaches yields individual advantages and disadvantages in certain fields of applications. Hence, the choice of the dedicated CFD approach strongly depends on the nature of the problem. Overall, these CFD methods provide effective measures to simulate a wide range of scenarios. Thus, CFD methods and corresponding simulation software are recognized as important tools in both, research and industry.

However, despite all these prospects, the simulation results are still typically not as reliable as physical measurements. This is mainly caused by the high numerical complexity of CFD, where the accuracy of the results are normally directly related to the available computer resources. As such, there is an inherent trade-off between accuracy of the results and computational time. The actual shortcomings that cause the huge numerical complexity thereby strongly depend on the dedicated CFD approach. More precisely, the following aspects have an impact in this regard:

- **Grid-based methods**: In grid-based methods, the accuracy of the solution is directly tied to the resolution of the underlying grid. The equation system is only obtained in the grid points and corresponding numerical errors are reduced by utilizing a finer grid discretization. This is especially important when complex geometries need to be modeled (e.g. an aircraft or a car). Using a coarse discretization can lead to details of the geometry not being accurately captured. Thus, the key factor to obtain accurate results is to utilize a high resolution grid. However, employing straightforward grid discretization techniques frequently results in a huge number of grid points. This leads to severe shortcomings of corresponding solutions for both, the memory consumption as well as the computational time.
- **Volumetric decomposition methods**: While volumetric decomposition methods allow to use smaller and larger volumes as compared to standard grid-based methods, these methods yield severe shortcomings in scenarios where moving objects are involved. In fact, every time the input objects are rotated, the topology of the object is changed which leads to different volumetric decompositions. Hence, the volumetric decomposition needs to be applied frequently throughout

the simulation. Since this introduces severe computational overheads, corresponding simulations utilizing volumetric decompositions frequently require weeks of simulation time.

- **Particle-based methods**: Particle-based methods yield strong capabilities in certain aspects of fluid flow modeling, such as in modeling of multi-phase flows (e.g. air and liquid at the same time). However, for modeling real world scenarios, frequently several hundred millions of particles need to be considered. This sheer number of particles is required to accurately capture scenarios such as drainage of liquid through tiny holes. The resulting numerical complexity eventually causes large computational demands which can lead to extremely long computational times.

Due to these shortcomings, these methods are frequently still not applicable to model certain real world scenarios. Complex applications involving huge input data can often only be modeled by employing simplifications to reduce the numerical complexity. Corresponding simulations then still take weeks of computational time—even on dedicated computer clusters. Because of that, some of the main research and development efforts of CFD emphasize on coping with these complexities.

This book sets the goal to improve on the aspects listed above, and to eventually make these CFD methods more practically relevant and applicable to real world phenomena. To that end, the main emphasis is on the optimization of the corresponding methods by employing *High Performance Computing* (HPC) techniques. The relevant aspects outlined above are investigated from an HPC perspective—yielding efficient approaches and methodologies allowing for coping with the huge complexities of the corresponding CFD methods. More precisely, the utilization of e.g. parallelism, optimized data structures, or powerful optimization algorithms for reducing complexities provide highly efficient approaches that significantly outperform current state of the art. Moreover, the proposed methodologies allow to consider more complex industrial cases, which could not be considered by methods available thus far. All these developments further contribute to the envisioned goal of making these CFD methods more practically relevant and applicable to real world scenarios.

In the following, the respective contributions are discussed for each of the CFD approaches reviewed above:

### Grid-Based Methods

Part II of this book presents a new grid discretization scheme which is based on so-called overset grids. The proposed scheme employs different levels of discretization, which allows to model the area of interest in a very fine resolution, whereas areas of less interest are modeled by a coarser grid. Following this scheme eventually allows to significantly reduce the number of required grid points by still maintaining the accuracy. The scheme proposed in this part differs to other overset grid schemes proposed thus far, as it specifically ensures that even very complex geometries are discretized by at least two layers of grid points, which is an essential aspect in a range of applications. More precisely:

- Chapter 3 provides an overview of grid-based methods, in particular the *Finite Difference Method* (FDM), which is considered in the remainder of this part of the book. Corresponding shortcomings related to the grid discretization are reviewed in detail, as well as how these shortcomings prevent the application of the FDM to dedicated applications in the industrial paint shop thus far.
- Chapter 4 aims to overcome the inherent drawbacks of the FDM caused by the grid discretization. To that end, a sophisticated grid discretization technique based on overset grids is proposed. The idea of this scheme is to have finer grids around the area of interest and coarser grids in areas of less interest. This technique eventually allows to efficiently model complex domains involving areas of varying interest and potential of error. Following this scheme finally allows to reduce the number of grid points by a factor of more than $1e^4$ as compared to a naive approach.

Overall, the overset grid scheme proposed in Part II of this book allows to significantly reduce the complexity of the grid discretization by maintaining the accuracy. This is confirmed by extensive experimental evaluations based on both, a numerical analysis and an industrial test scenario. In this industrial scenario, a dedicated application of the automotive paint shop which is used to apply corrosion protection on car bodies is considered. The methods proposed in this part of the book led to a novel simulation tool, which allows to accurately model this complex scenario in reasonable computational time.

**Volumetric Decomposition Methods**

Part III presents a framework which allows to compute the volumetric decomposition concurrently and, by that, significantly reduces the computational time. The basic idea of this scheme is to compute decompositions in parallel, rather than computing them sequentially for each discrete time step of a simulation. To this end, the proposed scheme employs parallelism on a *threading* level (i.e. on shared memory architectures) as well as on a *process* level (i.e. on distributed memory architectures). Moreover, the framework introduces two layers of parallelism by employing a hybrid shared and distributed memory approach. Dedicated workload distribution and memory optimization methods are presented, which result in significant speedups. More precisely:

- Chapter 5 reviews a volumetric decomposition method, which is based on so-called *reeb graphs*. Corresponding shortcomings are again discussed by means of applications of the automotive paint shop, which frequently led to weeks of computational time using state of the art decomposition methods used thus far.
- Chapter 6 presents a parallel scheme on a *threading* level. To this end, the scheme employs two levels of parallelism, where the outer layer enables the parallel computation of respective volumetric decompositions, while the inner layer allows for parallelism inside the actual decomposition methodology. Following

this scheme allows to significantly reduce the computational time, which is shown by means of a typical use case from the automotive industry.

- Chapter 7 aims to extend the shared memory parallel scheme for distributed memory architectures. To that end, major aspects related to initial workload distribution, optimizations with respect to serial dependencies, memory management, and load balancing are considered. The correspondingly obtained methods are again evaluated by means of a typical use case in the automotive paint shop, which confirms the efficiency and scalability of the proposed methods.

Overall, the approaches proposed in Part III of the book significantly improve on the state of the art of existing volumetric decomposition methods. The achievements are again confirmed by extensive evaluations employing a dedicated use case of the automotive paint shop. It is shown that the computational time could be reduced from more than one week to less than 15 h—a significant improvement.

**Particle-Based Methods**
Eventually, Part IV of the book introduces novel optimization techniques for the well known particle-based method *Smoothed Particle Hydrodynamics* (SPH). The discrete particle formulation of physical quantities makes particle-based methods particularly suitable for parallel architectures. More precisely, the sheer number of independent per-particle computations makes them a promising method for the *General Purpose Computations on Graphics Processing Units* (GPGPU) technology. To this end, a dedicated multi-GPU architecture is proposed, which employs spatial subdivisions to partition the domain into individual subdomains. These subdomains are distributed to the corresponding GPUs and executed in parallel. More precisely, research on SPH using dedicated multi-GPU architectures conducted within the scope of this book led to the following accomplishments:

- Chapter 8 reviews the general idea of particle-based methods and the SPH method in particular. Corresponding shortcomings are discussed, and it is shown how these can be overcome by using dedicated *High Performance Computing* (HPC) techniques. Moreover, a detailed review on state of the art HPC techniques used in the field of particle-based modeling thus far is provided.
- Chapter 9 presents a dedicated multi-GPU architecture for large-scale industrial SPH simulations. To that end, an advanced load balancing scheme is proposed, which allows to dynamically balance the workload between distinct GPUs. For that purpose, the employed methodology frequently conducts a so-called *decomposition correction*, which adjusts the workload among the multi-GPU architecture. Moreover, dedicated memory handling schemes allow to reduce the synchronization overhead introduced by load balancing methods employed thus far. The proposed scheme thereby addresses the inherent overhead introduced by multi-GPU architectures.
- Finally, Chap. 10 introduces dedicated optimization schemes for specific variants of SPH, which do not provide a straightforward parallelization scheme. Here, emphasis is on a promising variant called *Predictive-Corrective Incompressible SPH* (PCISPH), which typically allows to outperform other SPH variants by up

to one order of magnitude. However, PCISPH yields increased synchronization overhead, which thus far prevents it from exploiting HPC technologies. Due to that, PCISPH did not find a wide range of applications thus far. This book presents, for the first time, a PCISPH implementation that is suitable for multi-GPU architectures. To that end, dedicated load balancing and optimization techniques are presented that allow for efficient hardware utilization and, by that, allow to fully exploit the inherent advantages of PCISPH.

Overall, the methodologies proposed within Part IV of the book significantly improve the current state of the art of HPC techniques for particle-based methods. The efficiency and scalability of the proposed methods are confirmed by both, academic use cases as well as industrial scenarios. These industrial scenarios include different applications of the automotive paint shop. It is shown that the proposed methods allow to efficiently model even very complex industrial real world scenarios.

To keep this book self-contained, the next chapter of this first part of the book provides the background on CFD—including the fundamental equations and corresponding challenges. Moreover, an overview of HPC methods is provided, as well as a brief overview of industrial automotive paint shop applications to which the proposed methods are applied to. This serves as a basis for the remaining parts of this book, which cover the optimization of grid-based methods (Part II), volumetric decomposition methods (Part III), and particle-based methods (Part IV) as outlined above. Finally, Part V concludes this book.

# Chapter 2
# Background

## 2.1 Computational Fluid Dynamics

In this section, the basic concepts of *Computational Fluid Dynamics* (CFD) are reviewed. This includes a brief introduction to the governing equations, as well as corresponding discretization techniques.

### 2.1.1 Fundamentals

CFD deals with modeling and simulation of fluid (gas or liquid) flows. Such fluid flows are governed by *Partial Differential Equations* (PDEs) that describe the conservation laws for energy, mass, and momentum. In CFD, these PDEs are replaced by a range of algebraic equations, based on which an approximate solution is obtained by employing a computer-based solution [46].

To that end, CFD includes methods for simulations (i.e. predictions) of fluid flow scenarios. By that, CFD allows to model scenarios which are otherwise difficult or costly to measure in real world. It enables engineers to perform simulations in a very early stage of development, without the need to perform experiments using physical prototypes. Generally, CFD allows for predictions in a cheaper and faster manner as compared to physical experiments. Hence, CFD simulations are employed in a wide range of engineering disciplines in both, science and industry. Applications of CFD include, besides others:

- Simulation of the aerodynamic behavior of vehicles in the automotive industry [22].
- Simulation of arterial characteristics in the medical sector [105].
- Simulation of rotors or propellers of helicopters in aircraft industry [19].
- Simulation of heat exchangers of living environments [7].

© The Author(s), under exclusive license to Springer Nature Switzerland AG 2021
K. Verma, R. Wille, *High Performance Simulation for Industrial Paint Shop Applications*, https://doi.org/10.1007/978-3-030-71625-7_2

**Fig. 2.1**  General process of a CFD simulation

For a more complete overview on the field of applications of CFD, the reader is referred to [67, 88, 101].

Although CFD provides cost-effective measures to simulate a wide range of engineering problems, the simulation results are typically not as reliable as physical measurements. This is mainly caused by the input data of the simulation, which frequently involves imprecisions. Moreover, the accuracy of the result is tied to the resources of the available hardware. As CFD computations usually require powerful computer hardware, there is a frequent trade-off between accuracy and computational time. Overall, modeling complex fluid flow scenarios is a challenging and error-prone task that requires extensive experience to obtain valid results.

The general process of a CFD simulation is illustrated in Fig. 2.1. It fundamentally includes three parts: pre-processing, solving, and post-processing.

The pre-processing stage normally includes the preparation of the input data (i.e. the *meshing* of the data). Here, it is key that the input data accurately represents the topology of the objects to be simulated. A small error embedded in the input data may have large effects on the solving stage and may result into completely wrong simulation results. Hence, most of the manual effort is invested into preparation of the input data to avoid the propagation of such errors. Subsequent to pre-processing is the solving stage. Here, the governing equations are solved based on the discretization of the input data. This process is fully computer-resident and does not involve any manual interaction. Finally, in the post-processing stage the results are visualized and interpreted. Here, variables of interests are plotted and colour contours are visualized for further interpretation.

Nowadays the main research and development effort of CFD emphasizes on the solving stage, in which the fundamental equations are solved. In the following these governing equations are discussed in more detail.

## 2.1.2  Governing Equations

The basic governing equations of fluid flows have been known for more than 150 years and are referred to as the *Navier-Stokes equations* [13]. There exist several simplifications of these equations, depending on which problem is simulated and which physical effects can consequently be neglected. In general, the motion of a fluid particle is described by its mass, energy, and momentum. The corresponding *conservation of mass*, *conservation of energy*, and *conservation of momentum* constitute the Navier-Stokes equations  [13]. These principles basically state that mass, energy and momentum remain stable constants within a closed system.

In the following, the corresponding equation terms are introduced. However, to keep this section brief, the derivation of each of these equations is not covered. For a detailed treatment of these equations including their derivatives, the reader is referred to [13, 16, 86].

The conservation of mass (also referred to as the *continuity equation*) states that the mass within a closed system is conserved and can be written as

$$\frac{\partial \rho}{\partial t} + \rho \left( \nabla \cdot \boldsymbol{u} \right), \tag{2.1}$$

where $\rho$ is the density and $\boldsymbol{u}$ the local velocity. For incompressible fluids the change of density is zero, hence Eq. (2.1) can be written as

$$\nabla \cdot \boldsymbol{u} = 0. \tag{2.2}$$

The conservation of momentum is based on Newton's second law, which is defined as

$$F = ma, \tag{2.3}$$

where $F$ is the force, $m$ the mass and $a$ the acceleration.

When Eq. (2.3) is applied to a fluid particle, it can be written as

$$\frac{\partial \rho \boldsymbol{u}}{\partial t} + \rho (\boldsymbol{u} \cdot \nabla) \boldsymbol{u} = -\nabla p + \rho \boldsymbol{g} + \nabla \cdot \tau_{ij}, \tag{2.4}$$

where $\boldsymbol{u}$ is the local velocity, $\rho$ the density, $p$ the static pressure, $\boldsymbol{g}$ the gravity, and $\tau$ refers to the viscous stress tensors.

The conservation of energy states that the sum of work and heat flowing into the system from outside will result in increased energy in the system. The energy equation can be written as

$$p \left[ \frac{\partial h}{\partial t} + \nabla \cdot (h\boldsymbol{v}) \right] = -\frac{\partial p}{\partial t} + \nabla \cdot (k \nabla T) + \phi, \tag{2.5}$$

where $\frac{\partial h}{\partial t}$ defines the local change with time, $\nabla \cdot (h\boldsymbol{v})$ the convection term, $\frac{\partial p}{\partial t}$ the pressure, $\nabla \cdot (k \nabla T)$ the heat flux and $\phi$ the source term.

These equations of mass, momentum and energy constitute the Navier-Stokes equations, which are recognized as the fundamental equations of CFD.

In order to solve these equations, they are discretized in space by discrete sample points and then solved by algebra techniques. There are various discretization approaches, each of which yielding certain advantages and disadvantages. A well defined discretization approach is one of the key factors to obtain an accurate and fast solution. Hence, corresponding approaches are subject of ongoing

**Fig. 2.2** A simple 1D
domain. (**a**) Continuous 1D
domain. (**b**) Discretized 1D
domain

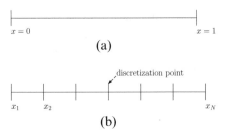

research activities. In the following, the most common discretization approaches are
reviewed.

### 2.1.3   Discretization Techniques

CFD is fundamentally based on the Navier-Stokes equations, which describe the
fluid motion. In order to solve these equations, the computational domain in
discretized in space. The Navier-Stokes equations are then computed on these
discretization points.

*Example 2.1* Consider a continuous 1D domain $0 < x < 1$ as shown in Fig. 2.2.
In a discretized domain, the space is only defined by $N$ discrete grid points $x =
x_1, x_2, \ldots, x_N$. The Navier-Stokes equations are only solved at the discrete points.

To this end, the input objects (e.g. a vehicle or an aircraft) and its surroundings
(e.g. a fluid or a solid object) need to be discretized by sample points that define the
topology of the space.

There are two main discretization techniques in fluid flow modeling: (1) grid-
based methods and (2) particle-based methods. The most common and widely used
approaches are based on grid-based methods. Here, the domain is discretized by a
grid which is fixed in space and time. Hence, it does not move nor deform as time
progresses. The fluid is constrained to stay with the grid. Some of the most common
grid-based methods are the *Finite Difference Method* (FDM, [85]), or the *Finite
Volume Method* (FVM, [23]).

The second approach relies on so-called particle-based methods. Here, the space
is discretized by particles that move freely in space. Each of the particles carries
corresponding physical quantities that define the underlying fluid. Some of the most
common particle-based methods are *Moving Particle Semi-implicit* (MPS, [103]),
*Finite Volume Particle Method* (FVPM, [37]) or *Smoothed Particle Hydrodynamics*
(SPH, [28]). These methods are all relatively new as compared to well-established
grid-based methods.

In the following, these grid-based and particle-based discretization techniques
are reviewed in more detail.

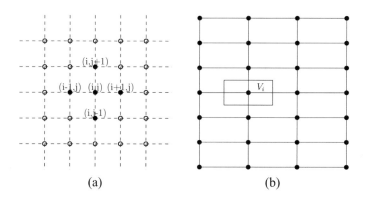

(a)                                                              (b)

**Fig. 2.3** Grid-based discretization techniques used in FDM and FVM. (**a**) Regular Cartesian grid discretization in 2D used in FDM. (**b**) Vertex-centered discretization in 2D used in FVM

### 2.1.3.1 Grid-Based Methods

The key idea of grid-based methods is to discretize the continuous domain into a computational grid. This grid does not move nor deform with time, but remains static throughout the whole simulation. The discrete grid points possess the physical quantities of the underlying fluid and the equations are correspondingly solved in these points. There are several different methods of discretization commonly used in CFD. Two of the most famous methods are the *Finite Difference Method* (FDM [85]) and the *Finite Volume Method* (FVM, [23]):

- *Finite Difference Method* (FDM): FDM is a well-studied technique, that typically employs structured grids as illustrated in 2D in Fig. 2.3a. Here, the domain is discretized into regular grid points constituting a *cell*. The cell size, i.e. the distance between the grid points is regular and fixed for the complete grid. Each of the grid points is identified by indices and neighboring points can be simply accessed by incrementing the corresponding index. In FDM, the numerical solution is obtained in these grid points. Hence, the accuracy of the solution is directly tied to the resolution of the grid. The main advantage of FDM is its simplicity and applicability to parallel computing. Its main disadvantage is the grid itself—in order to obtain an accurate solution a high resolutive grid is required which frequently brings this method to its limit. This shortcoming is subject for further optimization in Chap. 4.
- *Finite Volume Method* (FVM): In FVM the domain is discretized into cells in which the field quantities are evaluated in integral form. Compared to FDM, FVM cannot only utilize structured grids, but also unstructured grids (i.e. grids containing cells of different sizes). For that purpose, the domain is discretized into so-called *control volumes*. In each of these control volumes, an equation is obtained which is solved using numerical techniques. Figure 2.3b shows a vertex-centered discretization in 2D with a control volume $V_i$ used in FVM. Such mesh is also referred to as vertex-centered, as the numerical solution is stored in

each of the vertices (denoted by black dots). The advantage of FVM is a more natural conservation of mass and momentum, due to its integral form. Its main disadvantage is that for complex and irregular geometries, the preparation of the input mesh requires a lot of effort.

Overall, grid-based methods are well-studied techniques that find a large range of applications in both, science and industry. However, they also yield certain disadvantages, mainly in modeling multi-phase flows (e.g. liquid and air), or scenarios involving large deformations. This is an inherent drawback resulting from the grid itself. Underlying physical properties are only known in the discretization points. If this discretization is not fine enough, some details are frequently missed, e.g. the interface between liquid and air in multi-phase flows may not be accurately captured.

These scenarios can be more efficiently modeled using particle-based methods, as here such interfaces are inherently captured by the particles. The basics of these particle-based methods are discussed next.

### 2.1.3.2  Particle-Based Methods

In particle-based methods, the computational domain is discretized by particles. As against grid-based methods, these particles are not fixed, but move freely in time and space. Each of the particles carries certain physical quantities depending on which fluid or solid it is representing. Two of the most famous particle-based methods are *Smoothed Particle Hydrodynamics* (SPH, [28]) and *Moving Particle Semi-implicit* (MPS, [103]):

- *Smoothed Particle Hydrodynamics* (SPH): SPH is a particle-based fully Lagrangian method for fluid flow modeling. It was initially proposed by Gingold and Monagan [28] to simulate astrophysical phenomena. Nowadays, SPH is applied to a wide range of fluid flow scenarios, including multi-phase and free-surface flows, where the natural treatment of evolving interfaces makes it an enticing approach. The key idea of SPH is to discretize the domain by a set of particles. Based on these particles, a scalar quantity is computed by interpolating at a certain position by a weighted sum of neighboring contributions (i.e. neighboring particles) employing a gradient of a kernel function. SPH is covered in more detail in Chap. 8 and subject to further optimization in Chaps. 9 and 10.
- *Moving Particle Semi-implicit* (MPS); The MPS method is a particle-based method mainly applied to the simulation of incompressible free surface flows, initially proposed by Koshizuka and Oka [48]. In recent years, MPS has been applied to a wide range of engineering applications, such as mechanical engineering [35], structural engineering [11], or nuclear engineering [49]. Similar to SPH, in MPS the domain is discretized by particles, where scalar quantities are computed based on integral interpolates. However, as compared to SPH,

MPS does not employ a dedicated kernel function, but simply uses a weighted averaging to compute scalar quantities based on neighboring contributions.

Generally, particle-based methods yield strong capabilities in handling multiphase and free-surface flows, as well as scenarios involving large deformations. This results mainly from the particle-based representation, in which particles move freely according to physical laws and therefore there is no need to track any interfaces (i.e. the layer between two fluid phases).

However, despite these benefits, particle-based methods suffer from a huge numerical complexity. Especially for the simulation of real world phenomena, up to hundreds of millions of particles need to be considered—frequently resulting in weeks of simulation time. In order to overcome this shortcoming, *High Performance Computing* (HPC) techniques are leveraged. These techniques allow to significantly speedup such simulations, to eventually make particle-based methods more practically relevant and applicable to real world phenomena.

## 2.2   High Performance Computing

In this section, the basics of *High Performance Computing* (HPC) are introduced. This includes an overview of technologies and methods used in HPC, as well as corresponding limitations and applications.

### 2.2.1   Fundamentals

*High Performance Computing* (HPC) includes methods and technologies that emphasize on exploiting the computational power of supercomputers, or parallel computer architectures at large. For that purpose, HPC fundamentally employs methods for parallel processing to efficiently utilize computer hardware. To that end, HPC includes corresponding technologies for parallel processing and distributed computing, which eventually allows to process multiple tasks simultaneously on multiple processing units.

Overall, these parallel approaches frequently allow for significant speedup, as compared to straightforward serial implementations. The magnitude of the attainable speedup thereby strongly depends on the nature of the problem. Algorithms that exhibit a large degree of parallelism, i.e. individual tasks or data that can be processed simultaneously, generally allow for large speedups using HPC techniques. Ideally, of course, the achieved speedup is equal to the employed number of processing cores.

However, certain characteristics of parallel programs normally don't allow for such ideal scalability. This results from the fact that the parallel performance is limited by the amount of serial parts in the program. The equation describing the

attainable speedup as a balance between the parallel and serial part of a program is
known as Amdahl's law [1, 73]:

Let $p$ be the number of processing cores, $T_1$ the execution time on a single
processing core, and $T_p$ the execution time on $p$ processing cores. Then, the
maximum attainable speedup $S_p$ is defined as

$$S_p = \frac{T_1}{T_p} = p. \tag{2.6}$$

In practice, however, this ideal speedup can essentially never be reached due to
limiting factors such as inherently serial dependencies of an algorithm [73]. To
highlight the limiting factor of serialism in a program, let $f$ be the fraction of the
program which can be parallelized. The remaining fraction $1 - f$ is the inherently
serial part of the program and thus takes serial time ($T_s$):

$$T_s = (1 - f)T_1. \tag{2.7}$$

Then, the time spent on $p$ parallel processing cores is defined as

$$T_p = f\frac{T_1}{p}. \tag{2.8}$$

Based on that, the maximum attainable speedup is defined as

$$S_p = \frac{T_1}{T_s + T_p} = \frac{1}{1 - f + \frac{f}{p}}, \tag{2.9}$$

which is known as Amdahl's law [1, 73].

Thus, the inherent serial fraction of a program is a large limiting factor. Moreover, overheads introduced by communication and synchronization yield further
challenges to obtain optimal performance. To emphasize on this limitation, consider
again an ideal maximum speedup as defined by Eq. (2.6). Here, it is assumed that
no execution time spent on the communication between the individual processing
cores. However, in practice, there is a need for data exchange and communication
between the processing cores. This time to exchange data from local memory pools
across a communication network is referred to as the *latency* [73].

Let $T_c$ be the communication time required to exchange data. Then Eq. (2.6)
yields

$$S_p = \frac{T_1}{T_p + T_c} < p. \tag{2.10}$$

Hence, the communication time plays a key role in any parallelization strategy.
This time spent on communication is typically increasing as the number of process-

ing cores increases. Thus, employing more processing cores does not necessarily result into a larger speedup.

Overall, HPC is only a worthwhile technique for computationally intensive algorithms that exhibit a certain degree of parallelism. As such, HPC is applied to domains where large computing power can be leveraged, this includes, besides others,

- Chemical engineering,
- Aerospace studies,
- Geophysical studies,
- Computational Fluid Dynamics, as well as
- other engineering disciplines, such as mechanical engineering, energy engineering and more [70].

In the field of *Computational Fluid Dynamics* (CFD), HPC finds many different areas of applications due to its inherent compute-intensive nature as introduced in Sect. 2.1. In CFD, mainly a *divide and conquer* strategy is followed, where the original problem is divided into a set of subtasks. Each subtask is then individually executed on a processing core to exploit parallelism. For that purpose, methods for *domain decomposition* are utilized, that decompose the input mesh (i.e. the geometry describing the input object) into individual subdomains. Depending on the employed discretization strategy, frequent communication between these subdomains is required—yielding one of the main challenges in applying HPC technologies to CFD methods [42]. The three main approaches to facilitate parallelism in CFD are: (1) shared memory parallelization, (2) distributed memory parallelization, and, (3) GPGPU (*General-Purpose Computing on Graphics Processing Units*, i.e. the utilization of GPUs). While the former two approaches have been considered the most important parallelization schemes for many years, the latter become increasingly popular in recent years. Frequently, these approaches are also combined, yielding a *hybrid parallelization* approach including different levels of parallelism. In the following, these three approaches are reviewed in more detail.

## 2.2.2  Shared Memory Parallelism

In the shared memory model, processes share the same memory space to which they can read and write asynchronously. Figure 2.4 sketches the basic architecture used for the purpose of shared memory parallelization. The architecture is composed of multiple processing cores, each of which is able to access a globally shared memory. To that end, each processing core can operate independently to perform computations concurrently, whereas data is fetched from the global memory. Hence, there is no need to define any dedicated communication strategy between processes, as they are able to communicate via the shared address space.

The most common way to implement a shared memory model is to utilize threads, i.e. a single process fetches multiple threads which compute subroutines

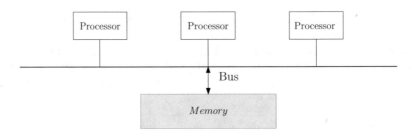

**Fig. 2.4** Shared memory parallel model

concurrently. One of the most widely-used techniques to implement such multi-threaded approach is to utilize OpenMP [18]. The OpenMP API provides platform independent compiler directives allowing to define the parallel regions. These pre-defined parallel regions are subdivided using a *divide and conquer* approach, resulting in individual data partitions. These individual data partitions are then processed by multiple threads concurrently.

The advantage of such shared memory model is its simplicity, as well as almost no overhead for communication, which is often rendered as the largest bottlenecks in parallel programs. One important disadvantage of this model is its limitation to single physical machines, i.e. it does not allow to parallelize across distinct machine boarders, and thus, provides limited scalability. As such, a shared memory approach is normally only applied to smaller scale problems where a single machine is sufficient. A shared memory approach is considered in Chap. 5, where it is used to parallelize a dedicated *decomposition method* for CFD computations. However, for larger scale problems, where a single machine is not sufficient anymore, typically a distributed memory model is leveraged.

### 2.2.3  Distributed Memory Parallelism

In the distributed memory model, multiple processes are used, each of which having its own individual local memory. Figure 2.5 sketches the basic architecture used for that purpose. The architecture is composed of multiple processing cores, whereas each processor has its own local memory. These processes can either reside on the same physical machine, but also across multiple machines that are connected via a network. Memory addresses of a local memories do not map to a global memory space, hence there is no concept of shared memory access as available in a shared memory model. Since each processing core uses an individual local memory, the processes are required to exchange data through an explicit communication protocol. The most common approach to implement such data communication is by employing *Message Passing Interface* (MPI, [15]). MPI provides standardized specifications to exchange data between processes. For that purpose, MPI defines an API which allows to exchange messages using e.g. send and receive directives.

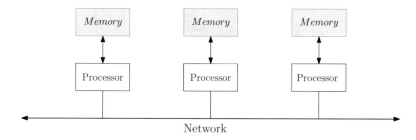

**Fig. 2.5**  Distributed memory parallel model

The advantage of a distributed memory model is its scalability, as it essentially allows to scale a parallel program to an arbitrary number of processing cores, even across individual physical machines. Its main disadvantage, however, is its requirement for data communication due to the utilization of local memory. These data transfers typically constitute the largest overhead in parallel programs, frequently limiting the parallel performance severely.

Such distributed memory approach is considered in Chap. 7, where it is combined with the shared memory approach of Chap. 5, resulting in a highly scalable parallel model for a dedicated *decomposition method* for CFD computations.

While a distributed memory method using CPUs has been the most important parallelization model for a few decades, in recent years, parallel models harnessing the computational power of GPUs have became increasingly important. In the following, the basics of such computations on GPUs are discussed.

## 2.2.4   General-Purpose Computing on Graphics Processing Units

Harnessing the computational power of GPUs for general-purpose computations is referred to as *General-Purpose Computing on Graphics Processing Units* (GPGPU). To that end, GPGPU includes methods for the use of GPUs for computations that are traditionally executed on CPUs [25, 26]. Originally, GPUs were designed for computer graphics, e.g. to accelerate graphics tasks such as image rendering. In recent years, however, GPUs are increasingly used as a platform for HPC, where they are used to harness the massively parallel architecture of GPUs.

Figure 2.6 illustrates the architectural differences between CPUs and GPUs. Here, in Fig. 2.6a it is shown that CPUs are composed of just a few ALUs (arithmetic logic units), whereas a large portion of transistors is dedicated to flow control and data caching. In contrast, as shown in Fig. 2.6b, GPUs usually contain significantly more ALUs (indicated by grey cubes in Fig. 2.6b) as compared to standard CPUs. Hence, significantly more transistors are devoted to data processing, rather than data caching or flow control [68].

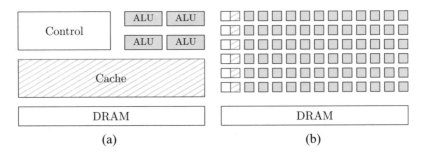

**Fig. 2.6** Schematic architectural comparison of CPUs and GPUs (based on [68]). (**a**) Schematic CPU architecture. (**b**) Schematic GPU architecture

**Table 2.1** Architectural specification comparison of an Intel Xeon CPU and a Nvidia Tesla GPU

|             | Intel Xeon E5-2670 CPU | Nvidia Tesla V100 GPU |
| ----------- | ---------------------- | --------------------- |
| Cores       | 8                      | 5100                  |
| Core clock  | 2.6 GHz                | 1.4 GHz               |
| Transistors | $2.3 \times 10^9$      | $21 \times 10^9$      |
| DRAM        | Flexible               | 32 GB                 |

To quantify these differences, Table 2.1 presents the architectural specifications of a state of the art CPU and GPU. While the CPU contains 8 processing cores, the GPU contains more than 5000. However, each of the GPU cores possess significantly slower clock speed (2.6 GHz compared to 1.4 GHz). Due to these architectural differences, GPUs are also referred to as *throughput devices*, i.e. devices that focus on executing many operations in parallel, rather than performing a single operation in minimal time. As such, harnessing the computational power GPUs can lead to severe speedups for problems that exhibit a large degree of parallelism.

However, the implementation of algorithms on GPUs is typically a highly non-trivial task. While the CPU provides straightforward and efficient random memory access patterns, the memory model of GPUs is divided into individual hierarchies and thus, is rendered much more restrictive. Hence, algorithms implemented for CPUs need to be fundamentally redesigned in order to make them applicable for computations on GPUs.

The most famous framework to facilitate such computations on GPUs is the CUDA (*Compute Unified Device Architecture*) framework by Nvidia [65]. It provides a software layer and APIs for direct access to virtual instruction set of GPUs, allowing to harness the computational power of GPUs.

In practice, GPGPU is frequently combined with distributed memory paralleliza-tion, where *message passing interfaces* are employed to utilize multiple GPUs, yielding a dedicated multi-GPU architecture. In such architectures, corresponding computations are not only parallelized on a single GPU, but distributed to multiple GPUs, and then executed in parallel. This not only increases the computational power, but also allows to overcome the inherent memory limitations of a single GPU, eventually allowing to consider more complex applications. However, this approach also introduces a large complexity related to memory handling, synchronization, or load balancing.

Overall, the utilization of GPUs and corresponding software frameworks is a highly non-trivial and cumbersome task which frequently requires a redesign of the whole application. However, harnessing the computational power of GPUs can results in significant speedups, especially for applications that exhibit a large degree of parallelism.

Due to these advantages, GPGPU finds application in many different areas, such as signal and image processing, computer graphics or Computational Fluid Dynamics (CFD). In the latter, it allows to deal with the inherent complexities of applications, where corresponding CFD simulations frequently require weeks of computational time using straightforward optimization techniques.

This is especially important in areas where complex data needs to be processed, which is frequently the case when CFD is applied to simulate complex real-world industrial problems. One of these areas, which is also further considered in this book, is the simulation of applications in industrial automotive paint shops. Here, typically complex geometries (e.g. full car bodies) and huge computational domains (e.g. large painting tanks) are involved, that frequently bring standard CFD techniques using straightforward HPC methods to their limits. In the following, an overview of applications of these automotive paint shops is provided, as well as corresponding challenges in the simulation of these applications which motivate the use of highly dedicated HPC techniques.

## 2.3  Automotive Paint Shop

In this section, the industrial automotive paint shop is discussed. This includes an overview of corresponding processes, as well as resulting problems and challenges in the paint shop.

### 2.3.1  Overview

The paint shop is one of the key processes in automotive manufacturing. It includes several different processes for cleansing, corrosion protection, and coating of car bodies or assemblies. The paint shop is considered as one of the major bottlenecks in the manufacturing, due to various time-consuming treatments involved in this stage [44].

From the perspective of economic indicators, the paint shop is rendered as a major cost of production, involving substantial material and energy costs [9].

A typical paint shop in automotive manufacturing roughly consists of the following processes [44]:

- Pre-treatment processes: The initial pre-treatment processes include operations for cleaning the bodies from dirt or dust. For that purpose, the bodies are washed in various stages using different liquids and cleansing mixtures. This typically involves liquid flows being injected by high pressure static nozzles. To that end, a set of nozzles is arranged in a certain spray pattern to ensure that the whole body is reached by the liquid injected by the nozzles. There are typically strict requirements attached to the cleanliness of the bodies, since defects in this stage can lead to severe shortcomings to the consecutive painting processes. Hence, the bodies are manually inspected after the cleaning stage to check for dirt or dust on the body surfaces [55].
- Corrosion protection processes: After the pre-treatment, corrosion protection is applied to the bodies. This mainly involves the so-called *Electrophoretic Deposition* (EPD, [6, 61]) process. Here, corrosion protection coatings are applied to the bodies, by moving them through a tank of liquid. Electrically charged paint particles then deposit on the bodies to form a corrosion paint layer on the surfaces. After EPD, the bodies are moved through an oven, where it is ensured that the paint layer dries and hence, sticks on the body surface.
- Painting processes: In the final process, the actual coatings are applied (i.e. certain colors). These coatings are typically applied in several layers, which includes: primer coat, base coat and clear coat. The primer coat mainly serves as a smoothing layer, whereas the base coat defines the actual color and visual impressions. The final layer of clear coat is transparent and mainly protects the coatings from environmental damages [102].

In each of these processes, strict quality requirements need to be met to ensure the desired quality in the final product. However, certain characteristics of the dedicated process frequently cause different errors in the corresponding processes. In the following, these challenges in the paint shop are discussed in more detail, which also motivates the utilization of dedicated simulations tools.

## 2.3.2 Challenges

The modern paint shop involves complex multi-stage processes, which are extremely energy and cost intensive. The paint shop is also the primary source for air emissions caused by different chemicals, as well as various waste caused by the coating itself, e.g. wasted paint through overspray, or cleansing mixtures [9]. Moreover, the quality of the coatings is often a criteria for product sales. In addition to that, many external factors such as temperature, humidity, or substrate composition can have a large impact on the quality of the coatings [9]. Hence, manufactures invest a lot of effort into optimizing corresponding paint shop processes to (1) reduce energy and material costs and (2) improve the coating quality.

For the latter, it is practically suited to perform the different coating processes on prototypes. Based on these prototypes, manufactures are able to assess the problematic areas on the car bodies and modify them to accordingly to subsequently overcome the certain problems. The following example illustrates this procedure:

*Example 2.2* An essential process in the paint shop is the application of corrosion protection by *Electrophoretic Deposition* (EPD). Here, a car body is dipped through a tank, and by electrically charged paint particles, the liquid is attached to the surface of the body. In EPD, it is desired that the entire surface of the car body is covered by the coating to essentially prevent corrosion. However, frequently the paint particles cannot reach to certain areas of the body (e.g. small cavities)—causing uncovered surfaces. Hence, EPD is performed on a prototype, based on which manufacturers are able to detect such problems (e.g. by cut-outs of surfaces) and conduct corresponding modifications (e.g. thrilling holes allowing paint particles to reach the problematic area).

Inherently, there are substantial costs attached to building such prototypes and performing corresponding painting processes. Moreover, prototypes can only be built in a very late development stage—implying that every modification made in the prototype stage requires to roll back to early design stages. This does not only cause immense costs, but may also lead to huge delays in the whole manufacturing process.

Therefore, there is a high demand for accurate simulation tools that allow to model corresponding paint shop processes. This not only drops the need for building costly prototypes, but allows to detect problematic areas in an early development stage. For that purpose, methods of *Computational Fluid Dynamics* (CFD) and corresponding simulation tools are utilized for the simulation of different paint shop processes. However, the inherent complexity of the automotive paint shop frequently brings standard CFD methods to its limits. This is mainly caused by the following characteristics of processes in the paint shop:

1. Complex physical behavior: Certain applications in the paint shop frequently require to consider complex physics, e.g. multi-phase fluid flows. This occasionally requires considering alternative and less-established CFD methods, e.g. particle-based methods, as against well studied grid-based methods.
2. Complex geometries of the input data: The automotive paint shop normally involves entire car bodies, which are geometrically extremely complex. Since an accurate representation of the geometry is one of the key factors for any CFD simulation to obtain precise results, a huge amount of input data that accurately represents the shape of the car bodies needs to be considered. This frequently leads to months of computational time, even on dedicated computer clusters.
3. Large computational domains: Paint shop applications usually require to consider large domains. E.g. the typical dimensions of a coating tank employed for the application of corrosion protection are in the range of $30 \times 5 \times 3$ m. Such

large computational domains result into huge numerical complexity—bringing standard optimization techniques to their limits.

Moreover, most painting processes are highly dynamic applications, for which different coatings are carried out by transient robots. These highly dynamic processes require to consider different movements (e.g. rotation and translation of objects) which introduces further complexity to the simulation. Based on all of this, simulation tools that accurately model paint shop applications are still subject of research nowadays [9]. Because of that, corresponding paint shop applications are further considered for experimental evaluations and benchmarks for the CFD and HPC methods proposed in this book.

# Part II
# Grid-Based Methods

This part of the book covers simulations based on grid-based methods. As reviewed in Sect. 2.1, there exist several different grid-based *Computational Fluid Dynamics* (CFD) methods, each of which yielding certain advantages and disadvantages. Here, simulations based on the *Finite Difference Method* (FDM, [85]) are considered. In FDM, the domain is discretized by a fixed regular grid and the corresponding fluid properties are computed on these fixed grid points. This grid does not move nor deform over time, but remains fix throughout the whole simulation. Based on this grid, a system of *Partial Differential Equations* (PDEs) is converted into a system of linear equations, which can be solved by techniques of matrix algebra. Some advantages of FDM are its numerical accuracy as well as robustness in a range of applications.

However, despite these benefits, one of the main disadvantages of FDM is the grid itself. More precisely, the fluid properties are constrained to only exist on the discrete grid points of the domain. Because of this, the accuracy of FDM is directly tied to the resolution of this discretization. Corresponding errors are reduced by increasing the number of discretization points. Hence, to provide realistic results, typically an extremely high resolutive grid discretization is required. Applying straightforward grid discretization techniques therefore frequently results in a huge number of grid points. This eventually leads to severe memory overheads, as well as increasing computational costs.

To overcome this shortcoming, it is required to employ more sophisticated grid discretization techniques. To that end, in this part of the book, a dedicated overset grid scheme is presented, which allows to overcome the inherent drawback of the huge number of grid points while maintaining the accuracy. The basic idea of this scheme is to have finer grids around the area of interest and coarser grids in areas of less interest. For that purpose, the proposed scheme employs multiple levels of discretizations, whereas in each level a different resolution is utilized. This scheme eventually allows to efficiently model complex domains involving areas of varying priority and potential of error. Moreover, this scheme allows to efficiently handle moving objects in the simulation. Using straightforward discretization techniques frequently requires to recompute the discretization based on the new position of the moving objects. The proposed overset scheme overcomes this by employing stationary grids for which only the boundaries are moved. The obtained values in these boundaries are then mapped onto the stationary grid, which remains fix throughout the whole simulation.

The application of the proposed methodology is investigated in the context of *Electrophoretic Deposition* (EPD, [6, 61]), an application in the automotive manufacturing which is well suited for the proposed scheme. EPD coating is one of the key processes in automotive manufacturing, where it is used to prevent cars from corrosion. Here, the coatings are applied to car assemblies or entire car bodies by moving them through a tank of liquid. Anodes are placed inside this coating tank and the car body is connected to the cathode. By applying a certain voltage profile on the anodes, the potential diffuses in the liquid inside the tank. The electrically charged paint particles start to deposit near the cathode which then leads to a successive increase of the paint film thickness on the car body. Since all of this involves substantial energy costs, there is a high demand to optimize this process by predicting the paint film thickness in an early development stage.

This process of EPD is especially suited for the proposed method, as it exhibits regions of varying interest. Typically, the most critical areas are around the car body (i.e. the cathode), whereas the remaining area of the tank is less critical. Hence, in the proposed simulation method, the overset scheme is specifically employed to model the region around the cathode (e.g. a BIW) in a very fine resolution, whereas the rest of the tank is modeled by a coarser grid. This eventually allows to efficiently model this application by discretizing the cathode with a resolution as small as 1 mm, which would lead to severe computational and memory costs using naive approaches. The scalability of the proposed scheme is shown by both, an academic and an industrial use case.

To set the context for the contributions provided in this part of the book, Chapter 3 provides the fundamentals of the proposed simulation method, as well as the process of EPD coatings. Then, Chap. 4 presents the details on the proposed simulation method including a detailed treatment of the overset grid scheme.

# Chapter 3
# Overview

## 3.1 Finite Difference Method

The *Finite Difference Method* (FDM) is a numerical method for solving *Partial Differential Equations* (PDEs). It is a well studied technique that has been utilized in a wide range of numerical analysis problems. In this section, the basics of FDM are reviewed. This includes the basic formulations as well as the grid discretization technique, which is recognized as one of the key aspects of FDM.

### 3.1.1 Formulation

The key idea of FDM relies on the conversion of continuous functions to their discretely sampled counterparts. By that, a system of PDEs is converted into a set of linear equations. This system of linear equations can then be solved by using ordinary matrix algebra techniques [32, 62].

Figure 3.1 illustrates this technique by employing a discretization of a continuous function $y = f(x)$. The numerical solution is obtained in the discretization points $x$ and $x + h$, where $h$ defines the distance between two discretization points (i.e. $h$ defines the resolution of the discretization).

By using such discretization technique, inherently an error between the numerical solution and an exact solution is obtained. This error is referred to as *discretization error* or *truncation error*. The error can be reduced by using a finer discretization, i.e. with smaller $h$, the error reduces.

To discuss this explicitly, consider the derivative of a smooth function $f$ at a point $x \in \mathbb{R}$. Then, the first derivative is defined as

$$f'(x) = \lim_{h \to 0} \frac{f(x + h) - f(x)}{h}. \tag{3.1}$$

© The Author(s), under exclusive license to Springer Nature Switzerland AG 2021
K. Verma, R. Wille, *High Performance Simulation for Industrial Paint Shop Applications*, https://doi.org/10.1007/978-3-030-71625-7_3

**Fig. 3.1** Discretization of a
continuous function

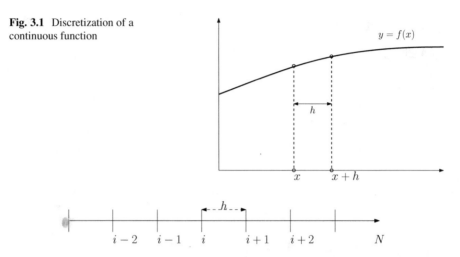

**Fig. 3.2** A 1D domain discretized among the x-axis. Discretization points are indicated by $i = i, \ldots, N$

Here, if $h$ tends towards 0, the right-hand side provides a good approximation of the derivative. Hence, the smaller the discretization, the more accurate the solution. The power of $h$ with which the error tends to 0, is referred to as the *order of accuracy of the difference approximation* [38]. It can be obtained by using a Taylor series expansion of $f(x + h)$. Equation (3.1) then yields

$$\frac{f(x + h) - f(x)}{h} = f'(x) + \frac{hf''(x)}{2!} + \frac{h^2 f'''(x)}{3!} + \cdots$$
$$= f'(x) + O(h^1). \tag{3.2}$$

Here, on the right-hand side the remaining of $f'(x)$ indicates the truncation error, also written as $O(h^1)$. If this error tends to 0, the approximation thus yields

$$f'(x) \approx \frac{f(x + h) - f(x)}{h}. \tag{3.3}$$

Based on this formulation, a set of PDEs is approximated at the discrete grid points to obtain a system of linear equations.

*Example 3.1* To explicitly apply this formulation, consider a 1D domain as sketched in Fig. 3.2, where the space is discretized among the x-axis. By this discretization, the continuum is sampled by $N$ discrete grid points $x_i$, were $i = i, \ldots, N$. $f_i$ indicates the value of the function $f(x)$ at the discrete points $x_i$. The distance between the grid points is constant and denoted by $h$. Applying the above formulations Eq. (3.2) and Eq. (3.3) at point $i$ then yields the following first derivative

$$f_i' = \frac{f_{i+1} - f_i}{h} - \frac{h f_i''}{2!} + \frac{h^2 f_i'''}{3!} + \cdots$$
$$= \frac{f_{i+1} - f_i}{h} + O(h^1).$$

Based this formulation, derivatives are approximated by employing neighboring function values using weighted computations. For that purpose, the completed computational domain is discretized by a structured grid. In each of the grid points, the local field quantities are stored and the approximations of the solution are computed as defined above.

The advantages of FDM are that it is relatively straightforward to implement and applicable for parallel computations. One of the main disadvantages, however, is the grid itself. In FDM, the solution is constrained to only exist in the grid points. Hence, the accuracy is directly tied to the resolution of the grid. Especially for complex shaped objects, an extremely high resolutive grid is required to capture all the geometrical details. Hence, in any FDM implementation, the grid discretization is recognized as one of the key aspects. This is discussed in more detail next.

## 3.1.2   Grid Discretization

They key idea of FDM is the discretization of the computational domain into discrete grid points. For that purpose, different spatial discretization methods have been developed in the past. One of the most common discretization techniques applied for FDM are structured grids.

Figure 3.3 illustrates such simple structured grid used in FDM (reduced to 2D for simplicity reasons). Each of the grid points is identified by the indices $(i, j)$ and the

**Fig. 3.3** Simple structured grid in 2D

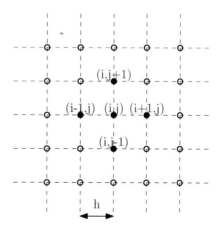

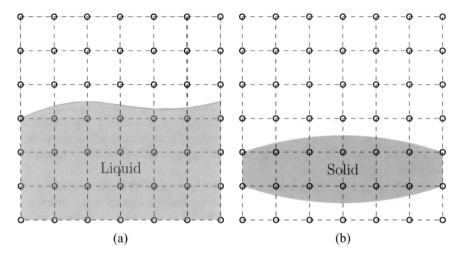

**Fig. 3.4** Discretizations employing regular grids in 2D. (**a**) Liquid-air domain decomposed by a regular grid in 2D. (**b**) Curved solid domain discretized by a regular grid in 2D

corresponding Cartesian coordinates $x_{(i,j)}$ and $y_{(i,j)}$. The cell size, i.e. the distance between to grid points is denoted by $h$.

The main advantage of such regular grid is its simplicity. Each line and row represents a linear address space, which renders addressing of certain grid points a trivial operation. Moreover, neighboring grid points can be obtained naturally by the neighboring grid points. Hence, there is no need for any computational expensive neighbor search technique.

However, since the numerical solution is constrained to only exist in the grid points, one disadvantage is that the accuracy of the solution is directly tied to the resolution of the grid. Hence, FDM has strong limitations in a range of fluid flow applications.

*Example 3.2* Consider a fluid domain as sketched in Fig. 3.4a. Here, the colored region represents liquid, whereas the white region denotes air. Using such discretization can easily lead to the conclusion that the complete surface of the fluid is defined in a single row, i.e. the shape of the surface is not accurately represented.

This behavior is one of the main shortcomings of regular grids. One way to address this is to employ a finer regular grid which allows to capture the interface (i.e. the region between liquid and air) more accurately. However, capturing such interfaces accurately can easily lead to extremely complex grids with hundreds of millions grid points. Obviously such complex system could result in months of computational time even on dedicated computer clusters.

Another shortcoming is the discretization of complex shaped geometries. Naturally, regular grids can be straightforwardly applied to regular objects such as cubes. However, when complex geometries are involved, regular grids yield severe shortcomings.

*Example 3.3* Consider a domain as sketched in Fig. 3.4b. Here, the curved solid object is discretized using a regular grid in 2D. Using such discretization can easily lead to the conclusion that the curvature of the object is not discretized accurately. In fact, the complete curvature is lost in the representation of the regular grid.

This example clearly shows that the resolution of the grid plays a key role in any FDM simulation. In order to provide realistic results, high resolution grids need to be employed. This frequently brings straightforward FDM implementations to their limits, specifically for industrial applications in which complex shaped objects need to be considered. One of such applications is the process of EPD coatings, in which typically whole car bodies need to be considered. Before this is explicitly discussed in Chap. 4, in the following, the basics of EPD coatings are introduced.

## 3.2 Electrophoretic Deposition Coatings

*Electrophoretic deposition* (EPD) is a process which is used in a wide range of applications, including the production of free-standing objects, laminated or graded materials, or infiltration of porous materials [91]. Moreover, EPD is used in the application of coatings, as widely used in automotive manufacturing.

The basic process of EPD coatings in automotive industry is sketched in Fig. 3.5. Here, a *Body In White* (BIW) is dipped through a tank that contains the liquid paint material. Anodes are placed at the side of the tank and a certain voltage profile is applied. Due to the current, the paint particles start to deposit near the cathode and the paint film thickness on the car body increases successively.

Overall, EPD coatings fundamentally includes three parts:

1. The object to be painted: Assemblies or BIWs which are connected to the cathode and dipped through the tank in a certain trajectory.

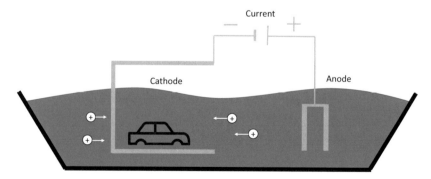

**Fig. 3.5** Basic process of EPD coating

2. The paint material: The liquid paint material which is stored inside the tank. It plays a key role, as the paint properties have a large influence on the distribution of the paint film thickness.
3. The anodes: The anodes are typically placed on the side of the tank and run with a certain voltage profile, usually in the range of 100–300 V. Sometimes anodes are also placed on the bottom of the tank or on the top (so-called *swimming anodes*), to account for low coating areas in these locations.

The primary electrochemical process occurring in EPD coating is the electrolysis of water [8]. The OH-ion is produced near the cathode and the paint particles begin to deposit near the cathode when more than a certain concentration of OH-ion is collected successively near the cathode. Since most of the paint particles adhere to the cathode, the resistance of the paint layer increases further over time. The corresponding increasing resistance of the built-up paint film causes the current density to decrease rapidly after initial high values.

Figure 3.6 exemplary shows the increase of the current density and deposited film thickness with time. A peak of the current density is observed in the beginning of the process, hence the film thickness is increasing rapidly. After initial high values, the current density decreases and the thickness deposition decreases accordingly.

This whole EPD coating process typically takes about 2–3 min in total. The final paint film thickness obtained on the cathode is usually in the range of 10–30 μm.

However, in order to properly protect the car body from corrosion, it is required that the whole surface area is covered by a certain paint film thickness. Frequently, areas inside the car bodies, e.g. in tiny cavities, are not properly exposed to the electric potential, hence no paint film thickness is achieved. To ensure that also such areas are properly painted, car manufactures frequently employ a physical prototype which allows to measure the paint film thickness. If the requirements are not met, changes to the process (e.g. an increase of the voltage) or modifications on the car body (e.g. drilling a hole to allow potential to diffuse inside) are conducted. Such a prototype can, of course, only be built in a very late development stage. Moreover, changes resulting from tests with the prototype often require a rollback to earlier development stages—frequently delaying the whole manufacturing process. Overall, this can result into substantial costs.

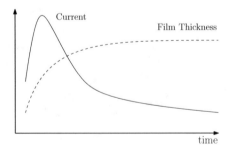

**Fig. 3.6** The sample current and the paint film thickness deposited over time. This process typically takes around 2–5 min

Thus, the simulation of this process is a great alternative, which not only renders expensive prototypes unnecessary, but also allows to detect defects at an early stage of development. Generally, grid-based methods are well suited to simulate such process, however the complexity of this application brings standard methods to their limits. However, the process of EPD exhibits different regions with areas of varying interest. While the area around the car body is the most critical, the remaining part of the paint tank is typically less critical. For that purpose, the following chapter introduces an overset grid scheme which utilizes this characteristic by employing different levels of discretizations to the simulation of EPD coatings. A thorough validation with both, analytical and industrial data confirms the accuracy and scalability of the proposed scheme.

# Chapter 4
# Simulation of Electrophoretic Deposition Coatings

The application of coatings by *Electrophoretic Deposition* (EPD, [6, 61]) is one of the key processes in automotive manufacturing. Here, it is used to apply coating that prevents the cars from corrosion. In order to properly protect the car body from corrosion, it is required that the whole surface area is covered by a certain paint film thickness. However, certain characteristics of this process frequently lead to areas with low paint film thickness, which is why there is a high demand for an accurate simulation tool.

In this chapter, a novel simulation method based on the so-called *Finite Difference Method* (FDM) is presented, which allows to simulate the paint film thickness in the EPD process (originally proposed in [96]). The proposed method considers exact physics and only requires surface meshes as input (as against other CFD methods which require volume meshes). To overcome the inherent drawback of FDM—the sheer number of required discretization points—a dedicated overset grid scheme is employed.

The proposed scheme uses a dedicated overset grid methodology which particularly ensures that even very thin metal sheets of the input objects are discretized by at least two layers. To that end, four grid levels are employed, which eventually allows for a discretization of whole car bodies with a resolution as small as $1mm$. Since the considered simulation arising in EPD exhibits multiple scales of behavior, with such an overset grid method, the differential equation is solved using hierarchy of discretizations. As a result, the computational costs and memory requirements are significantly reduced. The complexity of the moving nature of the application is overcome by employing stationary overset grids where only the boundaries are moved. Extensive experimental evaluations based on both, a numerical analysis and an industrial test scenario, confirm the accuracy and scalability of the proposed methods.

The remainder of this chapter is structured as follows: The next section provides the background on EPD coatings as well as the underlying formulations. Afterwards, Sect. 4.2 discusses the general idea of the proposed simulation method which is then covered in detail in Sect. 4.3. Finally, Sect. 4.4 summarizes the obtained results from our experimental evaluations before the chapter is concluded in Sect. 4.5.

## 4.1   Background

In this section, the background on the numerical modeling of the paint film thickness in EPD coatings is provided. This includes a comprehensive description of the underlying formulations, as well as a detailed treatment of the state of the art, as discussed next.

### 4.1.1   State of the Art

Unfortunately, rather few works have explored the simulation of paint film thickness in EPD coatings thus far. In [63] a paint resistance and deposition models for numerical simulation are discussed. In [80] a new non-linear film growth model for EPD coating simulation is presented, whereas the numerical solution is based on a finite element analysis. All these existing solutions rely on mesh-based *Computational Fluid Dynamics* (CFD, [2]) methods such as *Finite Volume Methods* (FVM, [98]), or *Finite Element Methods* (FEM, [82]). These methods are typically applied to simulate a large number of small volumes like meshes composed of tetrahedra or hexahedra [3, 87]. Only in [63], Boundary Element Method (BEM) is considered as the numerical solver, which requires a surface mesh. However, the solution matrix resulting from the BEM formulation is unsymmetrical and fully populated with non-zero elements, which can cause serious problems in memory storage and solution time, when the size of the problem is increased. Overall, simulating large objects such as entire car bodies frequently brings existing methods to their limits and, hence, typically requires significantly large computation times (even on dedicated HPC clusters).

### 4.1.2   Formulation

The basic formulation of EPD mainly depends on Faraday's laws of electrolysis, which is governed by

$$\Delta d = \frac{M}{\rho z F} j \Delta t, \tag{4.1}$$

where $d$ is the deposited paint thickness, $\rho$ is the specific weight, $M$ is the average weight of the deposited paint, $j$ is the magnitude of the local cathodic current density, $t$ is the time variable, and, $F$ is the Faraday's constant. The factor $a = \frac{M}{\rho z F}$ is also referred to as the *electrochemical equivalent*.

If the magnitude of the local cathodic current density is modeled as a function of space and time, the thickness can be computed by

$$d(x) = \frac{M}{\rho z F} \int_0^T j(x, t) dt. \tag{4.2}$$

By Ohm's law, the current density (vector) is obtained by

$$\mathbf{j} = -\kappa \nabla U, \tag{4.3}$$

where $\kappa$ is the conductivity and $U$ the electric potential. In the EPD process, the distribution of the electric potential $U = U(x, t)$ in the painting pool can be governed by a diffusion equation

$$\frac{\partial U}{\partial t} = \nabla \cdot (\kappa \nabla U). \tag{4.4}$$

Here, the conductivity of the liquid paint $\kappa$ is assumed to be a constant, which neither depends on space nor on a time variable. If a *quasistatic approximation* is assumed, Eq. (4.4) can be reduced to a Poisson equation

$$-\Delta U = 0. \tag{4.5}$$

These formulations of EPD provide the fundamentals for the proposed simulation method. This model can also be applied to other types of EPD, e.g. different substrates inside the tank. The corresponding physical properties of the considered material (e.g. dry film density, electro chemical equivalent, liquid electric conductivity, or solid conductivity) serve as an input to this model. Hence, the applied model is not only limited to a specific paint substrate.

Considering these equations, it can also be assessed that the fluid flow, i.e. the flow of the liquid inside the paint tank and the flow introduced by the moving object, is not considered in this approach. However, in Sect. 4.4, summarizing the conducted experiments, it is shown that even without the consideration of fluid flow, the simulation results are quite close to the values obtained by physical measurements.

The following section outlines how these formulations are considered in the proposed simulation method.

## 4.2   General Idea

In this section, the general idea of the proposed simulation method for the process of EPD coatings is outlined. To this end, at first the numerical modeling of the process is described. Afterwards, the grid discretization technique is discussed.

### 4.2.1   Numerical Modeling of EPD

For the numerical modeling of the EPD coatings process, the boundary conditions for Eq. (4.5) need to be established. As already described, there are three fundamental boundaries to consider: (1) the painting pool containing the liquid paint material, (2) the cathode, and, (3) the anodes. Hence, a multiple boundary condition problem needs to be solved. For simulating this process, the whole painting pool is modeled as the domain where the potential diffuses. The boundaries of the domain can be divided into three parts:

1. Anodes: A certain voltage profile is applied to the anodes which is considered as a Dirichlet boundary condition [10] governed by

$$U = U_{\text{anode}}, \tag{4.6}$$

   where $U_{anode}$ is the voltage given on the anodes.
2. Insulated wall: The side walls of the painting pool are considered as insulated parts. The insulated boundary including the air-liquid interface on top of the pool is considered as Neumann boundary condition governed by

$$\nabla U \cdot \mathbf{n} = 0. \tag{4.7}$$

3. Paint layer interface: The paint layer interface which is connected to the cathode is set as a Robin boundary condition [33]. This exactly represents the conductive state of the deposited paint layer between being conductive and insulated (i.e. the resistance increases with increasing paint film thickness). The interface is governed by

$$\nabla U \cdot \mathbf{n} + \gamma(d_n)U = 0, \tag{4.8}$$

   where the robin coefficient $\gamma(d) = \dfrac{\kappa_S}{d\kappa_F}$, and $\kappa_S$ and $\kappa_F$ are the conductivity of the solid paint and the conductivity of the liquid paint, respectively.

Taking these boundaries into account, the Poisson equation Eq. (4.5) needs to be solved. The solution yields the electric potential distribution inside the paint pool. Based on this solution and the electric potential on the cathode (paint layer interface) the paint film increase is computed based on Faraday's paint growth law Eq. (4.1).

There exists many partial differential equations (PDE) solving techniques, capable of solving such a mixed boundary condition problem. Each technique has its own certain characteristics and yields different advantages for specific types of PDEs. In this work, Finite Difference Method (FDM) has been chosen to solve for this mixed boundary condition problem, since (1) the mesh requirements for the input geometries are not very restrictive (e.g. it only requires a surface mesh rather than a 3D volume mesh), (2) it is well suited to solve such Poisson equation [64], (3) it is well studied and one of the most straightforward algorithms to solve for such PDEs.

The theory of FDM relies on discretization, converting a system of PDEs into a system of linear equations, which can be solved by matrix algebra techniques. The accuracy of FDM is therefore directly tied to the resolution of this discretization. Corresponding errors are reduced by increasing the number of discretization points. Especially in EPD coatings, the resolution of this discretization is key to obtain an accurate solution also inside tiny cavities of the car body.

The employed discretization strategy used for this purpose of the proposed simulation method is discussed next.

## 4.2.2   Grid Discretization

The key idea of FDM relies on the conversion of continuous functions to their discretely sampled counterparts. To obtain these discrete sample points, the whole problem domain is decomposed into a regular grid. To reduce the numerical error introduced by the discretization and in order to allow for the potential to also diffuse inside small cavities of the car body, the cell size $\Delta r$ of the regular grid must be chosen sufficiently small. In industrial use cases involving state of the art car bodies, hole sizes of such cavities can be as small as $<2$ mm. Hence, $\Delta r$ must be chosen $<2$ mm, which is not feasible when using a straightforward decomposition approach.

*Example 4.1* Consider the typical dimensions of the paint pool used in EPD coating, which are in the range of 30 m * 5 m * 3 m. Discretizing this domain into a regular grid with $\Delta r = 2$ mm would yield approximately $56 * 1e^9$ number of cells. Such extremely large number of cells obviously must be avoided in order to obtain an efficient solution.

In order to overcome this shortcoming of FDM, an overset grid approach is employed. The basic idea of overset grids is to have finer grids around the area of interest (i.e. the cathode) and coarser grids in areas of least interest and less error potential. The basic idea of a simple 2D overset grid using two levels is sketched in Fig. 4.1. Based on the solution computed on the coarser grid in level $\Omega^{2h}$, the obtained values are mapped and interpolated onto the finer grid level $\Omega^h$. Subsequently, the solution is obtained in the fine discretization points in level $\Omega^h$.

In the method used in this work, an overset grid approach using four levels is employed:

**Fig. 4.1** A simple two-level
overset grid approach in 2D

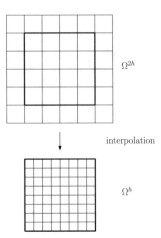

1. The outer level $\Omega^{16h}$: In this level the complete domain of the paint tank is modeled in a very coarse resolution.
2. The inner level $\Omega^{8h}$: Here, a bounding box around the cathode is created which defines the dimensions of this level.
3. The cathode level grid $\Omega^{2h}$: In this level, a grid only around the surface area around the cathode is created.
4. The thickness level grid $\Omega^{h}$: Here, only the cathode is discretized and the paint file thickness is computed based on the solution obtained in $\Omega^{2h}$.

By employing this overset grid approach, even small cell sizes of $\Delta r < 1\,\text{mm}$ are rendered feasible. This eventually allows for the proposed simulation method to accurately simulate the increase paint film thickness also inside small cavities of car bodies.

The details on these proposed simulation methodologies are discussed in the following section.

## 4.3 Simulation of EPD Coatings

In this section, the proposed simulation method for the process of EPD coatings is described. To this end, at first the detailed methodology of the solution is outlined. Afterwards, details on the overset grid representation are covered.

### 4.3.1 Implementation of Numerical Model

The numerical model as described in Sect. 4.1 is solved by first developing a discretization for the whole physical domain. To this end, the paint tank, anodes

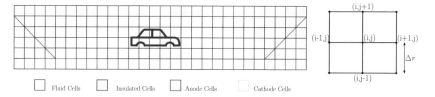

**Fig. 4.2** Simple grid representation of the domain. Reduced to 2D for simplicity reasons

---

**Algorithm 1** Compute thickness

---

1: Set: $U_0 = 0, d_0$. (Here $d_0 > 0$ is a starting thickness.)
2: **for** $n \in \{0, \ldots, N\}$ **do**
3:    Update time step $t_{n+1}$.
4:    Solve for $U_{n+1}$ according to the boundary conditions as defined in Sect. 4.2.1.
5:    Calculate current density $j_{n+1}$ according to Ohm's law.
6:    Update the paint layer thickness $d_{n+1}$ according to Faraday's law.
7: **end for**

---

and the cathode are at first discretized by axis aligned regular grid points with a cell size $\Delta r$. Based on this, the whole domain is decomposed into a regular rectangular grid with the same cell size $\Delta r$. The underlying grid cells are marked as either Anode, Fluid, Insulated Wall, or Cathode points. A simplified decomposed domain into a 2D rectangular regular grid is sketched in Fig. 4.2. Here, each cell of the grid is marked according to the underlying boundary. The cell sizes for the grid a regular, using a cell size of $\Delta r$.

Having the problem domain decomposed into these grid points, a numerical solution can be obtained. To that end, the potential distribution inside the domain is computed. Using the potential on the paint layer interface at the cathode cells, the deposited thickness values are computed based on Faraday's paint growth law employing Algorithm 1.

Here, at first the updated potential $U_{n+1}$ is obtained for each of the four boundaries by using the old paint layer thickness $d_n$ (see line 4). Based on the updated potential, the current density $j_{n+1}$ at the paint layer interface is computed (see line 5). Finally, the paint layer thickness $d_{n+1}$ is updated.

As already discussed, in EPD coatings the cathode is typically moving through the paint tank. That suggests that for each step $n \in \{0, \ldots, N\}$, the corresponding boundaries need to be updated. To that end, the discretized cathode paints are transformed according to the trajectory and the underlying grid cells are updated accordingly. This yields an algorithm as sketched in Algorithm 2.

Here, at first the domain is decomposed into the rectangular grid (see line 2). Then, in each time step the cathode points are transformed (see line 4) based on the moving trajectory of the process. For that purpose, each cathode point is at first rotated and, then, translated to the new position based on the current time step.

---

**Algorithm 2** EPD simulation

---

1: Initialize: $n := 0$, $t_0 = 0$, $U_0 = 0$, $d_0 = 0$
2: Compute Rectangular Grid
3: **for** $n \in \{0, \ldots, N\}$ **do**
4:      Move Cathode Points
5:      Mark Grid Cells
6:      Solve for updated potential $U_{n+1}$
7:      Calculate current density $j_{n+1}$
8:      Update the paint layer thickness $d_{n+1}$
9: **end for**

---

According to the new cathode point positioning, the corresponding grid cells are updated, i.e. the underlying grid cells are marked as cathode cells (see line 5). Now the updated potential $U_{n+1}$ inside the domain is solved (line 6), based on which the current density $j_{n+1}$ is obtained (line 7). Finally, the paint layer thickness $d_{n+1}$ is computed (see line 8).

As already discussed, in order to obtain an accurate solution for the paint layer thickness also inside the cathode, an appropriate resolution for the cell size $\Delta r$ has to be used. To still allow for an efficient solution in appropriate runtime, an overset grid approach is employed as discussed next.

## 4.3.2   Overset Grid Implementation

The proposed overset grid approach used in this work is based on four levels in total. To that end, these four levels are defined as follows:

1. Grid $\Omega^{16h}$: In this level, the complete domain including insulated wall, anodes, fluid and cathode is modeled.
2. Grid $\Omega^{8h}$: Here, a bounding box around the cathode is created and a new grid based on the bounding box dimensions is generated.
3. Grid $\Omega^{2h}$: In this level, only the area around the cathode is modeled.
4. Grid $\Omega^{h}$: In the finest level, the potential $U_{n+1}$ is not solved, but only the deposited paint layer $d_{n+1}$ is computed based on the potential obtained in $\Omega^{2h}$, since the gradient of potential at this level of discretization is negligible.

In the following, the rationale and details of each level are described. Afterwards, the resulting overall algorithm is presented.

### 4.3.2.1   Grid $\Omega^{16h}$

The movement of the cathode through the paint tank introduces additional complexity, since due to the new positioning corresponding grids need to be re-computed. To avoid this overhead, in the proposed approach, the cathode is only moved in $\Omega^{16h}$, whereas in all other levels a stationary grid is employed and only the boundary conditions are updated. To that end, in $\Omega^{16h}$ the complete domain including insulated wall, anodes, fluid and cathode is modeled in a very coarse resolution.

*Example 4.2* This procedure is sketched in 2D in Fig. 4.2. Here, the whole domain including all boundaries is modeled. The cathode body is moved according to the trajectory in each time step $n \in \{0, \ldots, N\}$ and the underlying cells of the grid are updated accordingly.

### 4.3.2.2   Grid $\Omega^{8h}$

Based on the current positioning, a bounding box around the cathode is computed which defines the boundaries for $\Omega^{8h}$. In every time step $n \in \{0, \ldots, N\}$, the cathode body in $\Omega^{16h}$ is moved to its corresponding position whereas in $\Omega^{8h}$ only the bounding box is moved. After solving for the potential distribution $U_{n+1}^{16h}$ in $\Omega^{16h}$, the corresponding potential is mapped onto the bounding box level $\Omega^{8h}$. Based on these obtained potential values on the boundary of $\Omega^{8h}$, the remaining finer grid points are linearly interpolated. The corresponding potential distribution $U_{n+1}^{8h}$ in level $\Omega^{8h}$ is then computed based on the interpolated potential values.

*Example 4.3* This procedure is sketched in 2D Fig. 4.3. Here, in $\Omega^{16h}$ 2 cathode positions are exemplary shown in step $n_t$ and $n_{t+m}$. The red dashed box denotes the bounding box around the cathode. In every time step $n \in \{0, \ldots, N\}$, the potential distribution $U_{n+1}^{16h}$ is mapped onto the bounding box of level $\Omega^{8h}$. The black crosses denote the known potential values computed in $\Omega^{16h}$.

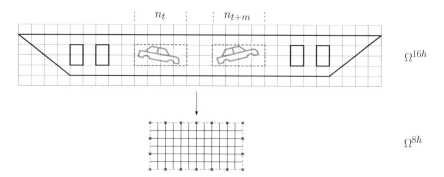

**Fig. 4.3**  Grid representation $\Omega^{16h}$ and $\Omega^{8h}$. In $\Omega^{16h}$ the cathode is moving, whereas in $\Omega^{8h}$ only the bounding box is moved and the obtained potential distribution $U_{n+1}^{16h}$ is mapped onto the stationary grid of $\Omega^{8h}$

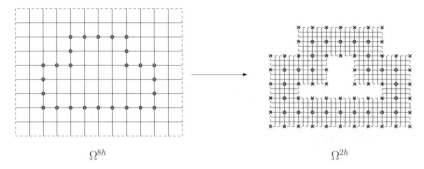

**Fig. 4.4** 2D grid representation of $\Omega^{8h}$ and $\Omega^{2h}$. In $\Omega^{2h}$ a grid only around the cathode cells (denoted by black circles) is generated

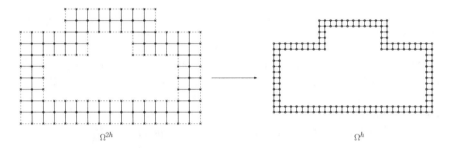

**Fig. 4.5** 2D grid representation of $\Omega^{2h}$ and $\Omega^{h}$. In $\Omega^{h}$ every cathode cell (denoted by black circles) of $\Omega^{2h}$ is split into four cathode cells (in 2D)

### 4.3.2.3  Grid $\Omega^{2h}$

In the following level $\Omega^{2h}$, a regular grid only around the cathode cells of $\Omega^{8h}$ is generated, i.e. inside and outside cells which are no cathode cells are neglected. By employing such grid, a massive amount of grid cells in both, inside and outside of the cathode body can be neglected. Similar to the procedure in $\Omega^{16h}$, the potential distribution $U_{n+1}^{8h}$ obtained in $\Omega^{8h}$ is mapped onto $\Omega^{2h}$ and the remaining values are linearly interpolated.

*Example 4.4* This procedure is sketched in 2D Fig. 4.4. Here, in $\Omega^{8h}$ the cathode points are exemplary denoted by black circles. In $\Omega^{2h}$, a grid is computed around these cathode cells including an additional cell layer in order to ensure a fluid layer around the cathode cells.

### 4.3.2.4  Grid $\Omega^{h}$

In the final level $\Omega^{h}$, each cathode cell of $\Omega^{2h}$ is split into eight individual cells. This is shown in Fig. 4.5, where each cathode cell of $\Omega^{2h}$ is split into four cathode cells (in 2D). All remaining fluid or anode cells are neglected, since in level $\Omega^h$ the potential distribution $U_{n+1}^h$ is not computed. It is shown that the potential gradient close to the cathode cells in $\Omega^{2h}$ yield negligible fluctuation between neighboring cell points. Hence, only corresponding potential values of the upper level $\Omega^{2h}$ are mapped onto the grid. Moreover, splitting up every cathode cell ensures that even very thin metal sheets of the cathode body are at least discretized by two layers of grid cells. The employed procedure in $\Omega^h$ ensures that for every part in the cathode unique values for inside and outside parts are obtained.

*Example 4.5* This procedure is illustrated in Fig. 4.6. Here, the black circle denote the cathode cells of $\Omega^h$ and the red dashed lines with black crosses denote the known potential values in $\Omega^{2h}$. For each cathode cell, the corresponding fluid cell of $\Omega^{2h}$ is computed, as indicated by the black arrows. Based on this potential, the final deposited paint layer $d_{n+1}$ is obtained in $\Omega^h$.

### 4.3.2.5   Discussion and Resulting Overall Algorithm

The presented methodology yields two major advantages: (1) it allows to significantly reduce the complexity of the whole simulation with minimum overhead introduced by overset grids, since corresponding grids only need to be computed once and, (2) the potential distribution $U_{n+1}^h$ does not need to be computed in the finest level $\Omega^h$, which results in even less computational work.

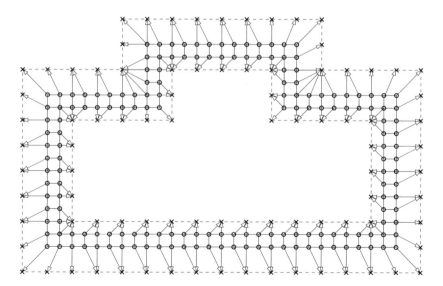

**Fig. 4.6** 2D grid representation of $\Omega^h$. The black circles denote the cathode cells, the red dashed line with black crosses denote the potential values of level $\Omega^{2h}$

---

**Algorithm 3** Overset grid solution

---
1: Compute Grid $\Omega^{16h}$, $\Omega^{8h}$, $\Omega^{2h}$, $\Omega^{h}$
2: **for** $n \in \{0, \dots, N\}$ **do**
3:      Move Cathode Points in $\Omega^{16h}$
4:      Mark Grid Cells in $\Omega^{16h}$
5:      Move Bounding Box of $\Omega^{8h}$
6:      Solve for $U_{n+1}^{16h}$ in $\Omega^{16h}$
7:      Map $U_{n+1}^{16h}$ and interpolate on $\Omega^{8h}$
8:      Solve for $U_{n+1}^{8h}$ in $\Omega^{8h}$
9:      Map $U_{n+1}^{8h}$ and interpolate on $\Omega^{2h}$
10:     Solve for $U_{n+1}^{2h}$ in $\Omega^{2h}$
11:     Map $U_{n+1}^{2h}$ and interpolate on $\Omega^{h}$
12:     Update $d_{n+1}$ in $\Omega^{h}$
13: **end for**

---

The proposed multi-level overset grid scheme also yields advantages over adaptive FDM grid approaches, e.g. used in [69]. The moving nature of this application typically results into increased workload for standard adaptive approaches, as typically the adaptive grids need to be recomputed when moving objects are involved. In the approach used in this work, the corresponding grids only need to be computed once in the beginning of the simulation and further remain constant. Moreover, the present approach guarantees that even very thin metal sheets are always discretized by at least two layers. This is essential for this application, since inherently the deposited paint film thickness differs largely between inside and outside regions.

The overall resulting algorithm of the overset grid approach is sketched in Algorithm 3. At first, the corresponding grids for all individual levels are generated (see line 1). Here, grid $\Omega^{16h}$ is created to represent the full domain. Then, $\Omega^{8h}$ is created based on the bounding box of the car. $\Omega^{2h}$ is created only around the cathode cells of $\Omega^{8h}$ and, finally, in $\Omega^{h}$ each cathode point of $\Omega^{2h}$ is split into eight individual cells. Then, in each time step $n \in \{0, \dots, N\}$ the cathode points are moved and the corresponding grid points are updated in $\Omega^{16h}$ (line 3–4). Based on the cathode positioning in $\Omega^{16h}$, the bounding box of $\Omega^{8h}$ is moved accordingly (see line 5). Next, the potential distribution $U_{n+1}$ is computed in all individual levels (line 6–10). Finally, the deposited paint film thickness $d_{n+1}$ is computed in the finest level $\Omega^{h}$ by considering the obtained potential distribution of $\Omega^{2h}$ (line 11–12).

## 4.4   Experimental Evaluations

In order to evaluate the accuracy and scalability of the proposed methods, experimental evaluations have been conducted which are summarized in this section.

More precisely, first an analytical study employing a simple (representative) object has been conducted. By means of this object, an analytical solution is derived which is used as a baseline to determine the numerical error introduced by the grid discretization. To that end, the result of the analytical solution is compared against the result of the simulation—showing the accuracy of the proposed method.

Afterwards, an industrial test case is used in which paint film thickness values are physically measured on a prototype. A range of measurement locations are defined, in which the measured paint film thickness values are compared against the values obtained in the simulation. Besides that, the performance of the proposed simulation method is evaluated by means of various simulations using a range of different cell sizes $\Delta r$—showing the scalability of the proposed method.

### *4.4.1 Validation with Analytical Data*

The employed grid discretization inherently introduces an abstraction which eventually leads to a numerical error in the solution. To evaluate the accuracy of the proposed methods, an analytical study using a very simple input is conducted. This input consists of a cubic domain (representing the paint tank) in which a cube is placed (representing the cathode, i.e. the car body). The cube is fixed inside the domain and does not move with time. Figure 4.7 illustrates this case inside the cubic box. Based on these simple objects, an analytical solution is derived. In this analytical study, the Poisson equation needs to be solved:

$$-\Delta U(\mathbf{x}) = 0, \qquad \mathbf{x} \in \Omega,$$

$$\nabla U(\mathbf{x}) \cdot \mathbf{n}(\mathbf{x}) + \gamma(\mathbf{x})U(\mathbf{x}) = 0, \qquad \mathbf{x} \in \Gamma_1,$$

$$U(\mathbf{x}) = U_{\text{anode}}, \qquad \mathbf{x} \in \Gamma_2,$$

$$\nabla U(\mathbf{x}) = 0, \qquad \mathbf{x} \in \Gamma_3,$$

**Fig. 4.7** An object with a fichera corner inside a cubic box

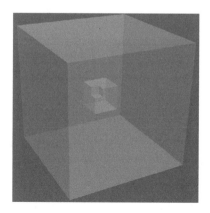

where $\mathbf{n}(\mathbf{x})$ is the unit normal vector pointing to the outside of $\Omega$. $\Omega$ is a 3D domain representing the paint path pool and $U$ is the electric potential distribution in the domain $\Omega$. $\Gamma_1$ represents the paint layer interface (normally the car body) connected to the cathode. $\Gamma_2$ represents the anodes and $\Gamma_3$ the insulated wall. $\gamma(\mathbf{x})$ will be updated with the film thickness in each time step and $U_{\text{anode}}$ is the given voltage profile on the anode.

In this test example, let $\Omega$ be a cubic box with an object $X$ cut out (also referred to as a *fichera corner*, i.e.

$$\Omega = [-1, 0] \times [-1, 0] \times [0, 1]\backslash X.$$

This represents an object $X$ in a cubic paint bath pool. Let $\Gamma = \bigcup\limits_{i=1}^{3} \Gamma_i$ be the boundary of $\Omega$. Here

$$\Gamma_1 = \partial X$$
$$\Gamma_2 = \{-1\} \times [-1, 0] \times [0, 1] \cup \{0\} \times [-1, 0] \times [0, 1],$$
$$\Gamma_3 = [-1, 0] \times \{-1\} \times [0, 1] \cup [-1, 0] \times \{0\} \times [0, 1]$$
$$\cup [-1, 0] \times [-1, 0] \times \{0\} \cup [-1, 0] \times [-1, 0] \times \{1\}.$$

In this example, the potential on the anodes $U_{\text{anode}}$ is given by

$$U_{\text{anode}} = \cos(\pi y)\cos(\pi z),$$

and $\gamma(\mathbf{x})$ is given by

$$\gamma(x, y, z) = \begin{cases} -\sqrt{2}\pi \tanh\left(\sqrt{2}\pi\left(x + \frac{1}{2}\right)\right), & \text{if } \mathbf{n}(\mathbf{x}) = (1, 0, 0)^T, \\ \sqrt{2}\pi \tanh\left(\sqrt{2}\pi\left(x + \frac{1}{2}\right)\right), & \text{if } \mathbf{n}(\mathbf{x}) = (-1, 0, 0)^T, \\ \pi \tan(\pi y), & \text{if } \mathbf{n}(\mathbf{x}) = (0, 1, 0)^T, \\ -\pi \tan(\pi y), & \text{if } \mathbf{n}(\mathbf{x}) = (0, -1, 0)^T, \\ \pi \tan(\pi z), & \text{if } \mathbf{n}(\mathbf{x}) = (0, 0, 1)^T, \\ -\pi \tan(\pi z), & \text{if } \mathbf{n}(\mathbf{x}) = (0, 0, -1)^T. \end{cases}$$

Based on this domain, the exact solution can be obtained by solving

$$U(\mathbf{x}) = \frac{\cos(\pi y)\cos(\pi z)\cosh\left(\sqrt{2}\pi\left(x + \frac{1}{2}\right)\right)}{\cosh\left(\sqrt{2}\pi\frac{1}{2}\right)}.$$

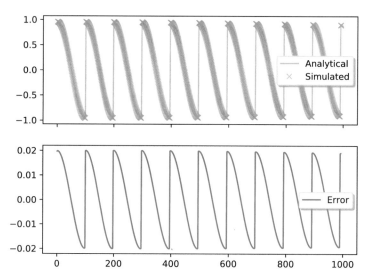

**Fig. 4.8** Comparison of analytical and simulated potential values on top and corresponding error on the bottom

In this test example, consider $X$ similar to be a so-called *fichera corner*, i.e. a cube with a cube cut out at a corner:

$$X = ([-0.6, -0.4] \times [-0.6, -0.4] \times [0.4, 0.6])$$

$$\backslash([-0.6, -0.533] \times [-0.6, -0.533] \times [0.533, 0.6])$$

Based on these equations, the analytical solution can be computed for the present case. By means of this test case, a comparison between the analytical solution and the simulation results is conducted.

Figure 4.8 shows the obtained results of the analytical solution and the simulation in a range of locations inside the domain. The plot on the top shows the comparison between the analytical and simulated result, whereas the plot on the bottom shows the corresponding error. It can be assessed that the simulated results are very close to the analytical solution, with an error of <2%. This confirms that the abstraction introduced by the grid discretization yields a negligible error as compared to an exact analytical solution.

## 4.4.2 Validation with Industrial Data

As an industrial test scenario, the coating of two truck cabins as shown in Fig. 4.9 is considered. Here, two cabins are simultaneously dipped into the coating tank. Once fully dipped in, the cabins are rotated 10° in positive and negative direction along

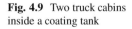

**Fig. 4.9** Two truck cabins
inside a coating tank

the y-axis. In the end of the process, the two cabins are simultaneously dipped out
again. This trajectory is obtained by actual drawings of the paint tank and, then,
defined as discrete time steps for the simulation method. In between the discrete
time steps, the movement is linearly interpolated. The anodes are placed close to
the side wall of the tank and run with a voltage program of up to 340 V. The whole
coating process takes 160 s in total.

In order to evaluate the accuracy of the proposed method, the paint film thickness
is mechanically measured by means of a prototype. To that end, 50 measurement
localizations are defined on the body in which the measured results are compared
against the simulation results. Here, measurement locations 1–12 are on the front,
13–26 are on the left side, 27–37 on the back side (i.e. in between the cabins), and
38–50 on the right side of the cabin.

The obtained results are visualized in Fig. 4.10. It is shown that the simulation
results are generally in good agreement with the measurements, with the discrep-
ancy being in the range of $\pm 2\,\mu$m. Considering that this test scenario comprises
complex industrial conditions, the results are rendered satisfying. Moreover, by
means of the simulation results, the problematic areas can easily be assessed. The
simulation clearly displays that the area in between the two cabins yield less paint
film thickness values with being less than $16\,\mu$m. Hence, the simulation results
would allow to optimize the process (e.g. increase the voltage profile for the anodes
close to the middle of the two cabins in Fig. 4.9) without the need of performing
measurements on a prototype.

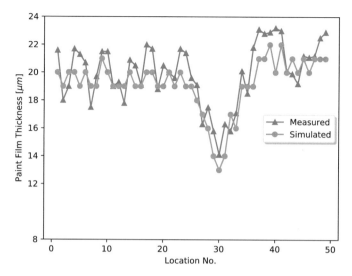

**Fig. 4.10** Comparison of measured and simulated paint film thickness values

**Table 4.1** Number of cells using different cell sizes $\Delta r$

| $\Delta r$ [mm] | $\Omega^{16h}$ | $\Omega^{8h}$ | $\Omega^{2h}$ | $\Omega^{h}$ | Sum | Naive |
|---|---|---|---|---|---|---|
| 8 | $102 \times 10^3$ | $89 \times 10^3$ | $1.5 \times 10^6$ | $1.1 \times 10^6$ | $2.8 \times 10^6$ | $370 \times 10^6$ |
| 6 | $236 \times 10^3$ | $184 \times 10^3$ | $2.9 \times 10^6$ | $2.2 \times 10^6$ | $5.5 \times 10^6$ | $927 \times 10^6$ |
| 3 | $1.8 \times 10^6$ | $1.1 \times 10^6$ | $13 \times 10^6$ | $10 \times 10^6$ | $27 \times 10^6$ | $7.4 \times 10^9$ |
| 2 | $5.9 \times 10^6$ | $3.5 \times 10^6$ | $33 \times 10^6$ | $24 \times 10^6$ | $67 \times 10^6$ | $18.5 \times 10^9$ |
| 1 | $46 \times 10^6$ | $25 \times 10^6$ | $153 \times 10^6$ | $103 \times 10^6$ | $282 \times 10^6$ | $1.5 \times 10^{11}$ |

## *4.4.3   Performance Discussion*

The accuracy of the proposed methodology is directly tied to the resolution of the grid discretization. Especially for industrial scenarios, the paint film thickness inside small cavities is crucial. Therefore, the applicability of the simulation method strongly depends on the resolution which is feasible in acceptable runtime. In order to evaluate this feasibility, a range of simulations using an industrial scenario is performed. This scenario is using a state of the art car body and a typical paint tank with dimensions of approximately 23 m * 3 m * 2.7 m.

Table 4.1 summarizes the resulting number of cells obtained for different resolutions. For each level the number of cells is listed, as well as the sum over all levels (i.e. the total number of cells). The leftmost column states the cell size used in the finest level $\Omega^h$. The rightmost column exemplary shows the number of cells employing a naive approach with only a single grid level (i.e. discretizing the whole domain with a single grid of a cell size as listed in the leftmost column). The results clearly show that the proposed overset grid approach improves significantly upon a naive approach. In the approach used in this work, the total number of cells

**Table 4.2** Runtime and
memory consumption using
different cell sizes $\Delta r$

| $\Delta r$ [mm] | Runtime [h] | RAM [GB] |
|---|---|---|
| 8 | 15 | 18 |
| 6 | 20 | 20 |
| 3 | 30 | 32 |
| 2 | 45 | 55 |
| 1 | 70 | 80 |

for a cell size $\Delta r = 1mm$ are reduced by a factor of more than $1e^4$ as against a naive approach.

Besides that, also the performance of the whole simulation is evaluated. For that purpose, the runtime and memory consumption of the whole simulation are evaluated on a workstation composed of an AMD Threadripper 2990WX, 3.0–4.2 GHz and 128 GB RAM on Ubuntu16.04. Table 4.2 summarizes the resulting runtime and memory consumption of the whole process. It is shown that even with a very small cell size of $\Delta r = 1$ mm, the results can be obtained in less than three days with reasonable memory consumption.

Overall, the performed test cases demonstrate that the proposed approach allows to efficiently simulate complex industrial scenarios using a fine resolution in reasonable runtime.

## 4.5   Summary

In this chapter, a novel simulation scheme for the simulation of EPD coating processes has been presented. The proposed scheme uses exact physics and employs an overset grid methodology to allow for a high resolution discretization. To that end, four grid levels are employed, in which the whole domain is modeled in a coarse resolution and the area around the object is discretized in a very fine resolution. The proposed approaches led to a simulation method, which accurately models industrial EPD coating processes in quick computation times. Experimental evaluations on both, an analytical study and an industrial scenario confirm the accuracy and scalability of the proposed methods. Future work includes investigations of the effect of fluid flow (i.e. the flow introduced of the object moved through the liquid) on the final deposited paint film thickness, as well es further optimization to the overset grid scheme.

# Part III
# Volumetric Decomposition Methods

This part of the book covers simulations based on so-called *volumetric decomposition methods*. Volumetric decomposition methods aim to use fewer and larger volume units as compared to standard CFD methods. For that purpose, a volumetric decomposition into so-called *flow volumes* has been introduced by Strodthoff et. al. [83] and initially applied to the commercial simulation tool *ALSIM* (from the German "Auslaufsimulation", i.e. drainage simulation). This volumetric decomposition allows to use triangular surface meshes, as against CFD, where typically volume meshes are required. Based on this input mesh, the object gets decomposed into flow volumes based on *critical vertices* of the mesh. As a result, a so-called *reeb graph* accurately represents the topology of the object. By using this approach, the number of volumetric units can be significantly reduced as compared to standard discretization approaches used in common CFD methods. This eventually allows to reduce both, the computational time and the memory consumption significantly.

However, despite these benefits, one of the main disadvantages of this method is that the volumetric decomposition has to be applied frequently throughout a simulation. In fact, every time the input object is rotated, the topology of the object is changed—leading to different volumetric decompositions. This is a crucial problem, since in most real world scenarios, moving objects that change their position frequently throughout the process are involved. Thus, the volumetric decomposition needs to be frequently applied for various rotation angles—leading to severe computational overheads.

To address this shortcoming, in this part of the book, a parallel framework is presented, which allows to compute corresponding volumetric decompositions concurrently and, by that, significantly reduce the computational time. To that end, a parallel scheme on a threading level for shared memory architectures, as well as on a process level for distributed memory architectures is introduced. The basic idea of this parallel scheme is to allow for independent parallel computations of each volumetric decomposition. Instead of computing the decomposition sequentially for every time step, $n$ decompositions are computed in parallel, where $n$ is the number of cores available on the system. Moreover, an *internal parallel layer* is introduced, which allows for data parallelism in each independent volumetric decomposition computation.

The application of this method is investigated in the context of *Electrophoretic Deposition* (EPD, [6, 61]). As reviewed in Chap. 3, in EPD car assemblies or entire car bodies are moved through a tank of liquid to apply corrosion protection. Here, volumetric decomposition methods are applied to simulate the emergence of air bubbles and liquid drainage during this process. These air bubbles frequently prevent the surface from being entirely covered by the coating, which can lead to corrosion.

This process of EPD is especially suited for the validation of the proposed framework, as here, the input object is dipped into the tank by a certain kinematic. In this transient and highly dynamic process, the object will be constantly rotated. Due to this inherent dynamic behavior, the volumetric decomposition needs to be applied frequently for various rotation angles. For that purpose, the proposed parallel scheme is specifically employed to compute these different rotation angles concurrently and hence, allow for a large degree of parallelism in the simulation of air bubbles and entrapped liquid in EPD. This eventually allows to significantly reduce the computational time from more than one week to less than one day. The scalability and efficiency of the proposed scheme is evaluated by a state of the art industrial use case scenario.

To set the context of the contributions provided in this part of the book, Chap. 5 provides the background of the volumetric decomposition method, including a detailed review on the fundamentals of this method, as well as the corresponding drawbacks. Chapter 6 discusses the parallel scheme for shared memory architectures, whereas the final chapter of this part of the book introduces the distributed memory parallel scheme.

# Chapter 5
# Overview

## 5.1 Fundamentals

The main idea of volumetric decomposition methods is to use fewer and larger volume units compared to standard CFD methods in order to reduce the computational complexity. For that purpose, the input model is a triangular surface mesh, as against CFD where typically volumetric meshes (consisting of e.g. tetrahedras) are widely used. The advantage of surfaces meshes over volumetric meshes is that significantly less elements need to be considered in order to describe the geometry of an object. However, such meshes only describe the surface of the object, whereas the volume (i.e. the inside region) is neglected. Such volumetric representation is essential in the field of CFD, where e.g. the fluid flow inside an object should be simulated.

Thus, in order to compute such volumetric representation of the object, a geometrical decomposition into so-called *flow volumes* has been introduced by Strodthoff et al. [83]. Here, flow volumes are defined as connected parts of a given triangulated solid, with the boundary consisting of triangles of the triangulated solid and parts of horizontal planes on top and bottom. To generate these flow volumes, the object is scanned for local minimums, maximums, and saddle points (also referred to as *critical vertices*) while sweeping from bottom to top. Each of these identified points ends the former flow volume and starts a new one.

*Example 5.1* Consider the object representation of a simple tube from Fig. 5.1a. This is geometrically decomposed by vertical cuts into flow volumes as illustrated in Fig. 5.1b. Each number denotes one identified flow volume.

Based on this volumetric decomposition, a graph is constructed which represents the topology of the object by flow volumes with their respective relations. The resulting graphs can be seen as so-called *reeb graphs* [21], which are originally a concept of Morse theory [56], where they are used to gather topological information. This graph representation describes the topology of the object, i.e. it describes the relation and connection between the individual flow volumes. This is essential in

© The Author(s), under exclusive license to Springer Nature Switzerland AG 2021
K. Verma, R. Wille, *High Performance Simulation for Industrial Paint Shop Applications*, https://doi.org/10.1007/978-3-030-71625-7_5

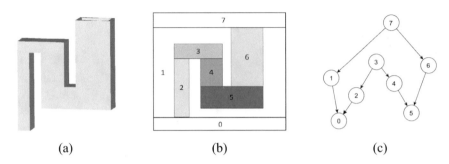

**Fig. 5.1** Geometric decomposition of a simple object. (**a**) Input object. (**b**) Flow volumes. (**c**) Resulting graph

CFD simulations, as this facilitates the computation of fluid flows inside the object by knowing possible flow paths of fluids.

*Example 5.2* Figure 5.1c shows the resulting graph, while the node numbers correspond to the identified flow volumes shown in Fig. 5.1b. By means of this graph, it is known that e.g. fluid of volume 3 can flow into volume 2 and volume 4, while fluid of volume 4 can only flow to volume 5.

Based on this volumetric decomposition, so-called *bottlenecks* need to be integrated into the graph. These are needed to consider liquid flows with respect to time. Without the consideration of time, liquids would touch every reachable surface of the object immediately. But of course, liquids need time to spread, especially when it goes through narrow channels. For that purpose, it is essential to detect all narrow channels of an object in order to limit the liquid flow speed. Such narrow channels are referred to as *bottlenecks*.

For the purpose of detecting bottlenecks, which are defined as the shortest loop round a narrow channel, an algorithm is employed which creates a distance field (using the Dijkstra algorithm) in order to find the shortest path between two nodes in a graph. Starting from one vertex of the mesh, its neighbors are added to the distance field, along with their neighbors, until the given circumference is reached. A bottleneck is found whenever the neighborhood of a newly inserted vertex fulfills a certain criterion.

As previously discussed, the flow volumes and their relations are represented by the reeb graph. Bottlenecks are found and represented as lists of vertices. For the simulation, their position with respect to the flow volumes is relevant. Therefore, the bottlenecks need to get integrated into the reeb graph.

There exist basically two types of bottlenecks, horizontal and vertical bottlenecks. Vertical bottlenecks can only appear inside of a volume, therefore a split of the volume is indispensable. Horizontal bottlenecks can either appear between two volumes or inside a volume. In case they appear inside a volume, the volume needs to be split, otherwise the respective node connection just needs to be marked as a bottleneck connection. This bottleneck integration significantly influences the flow volumes, and consequently the resulting reeb graph. Finally, the integration of these bottlenecks allows to limit the liquid flow through such narrow channels and hence, to more accurately capture the physical behavior of fluids.

Overall, this volumetric decomposition approach provides the main advantage compared to standard CFD methods that significantly less volumes need to be considered to define the topology of the object. However, it also yields a main disadvantage which is related to rotation. While in standard CFD approaches the volumetric input mesh can be simply rotated by applying a straightforward matrix multiplication, in volumetric decomposition methods, such operation is more complex. In fact, every time the object is rotated, the volumetric decomposition needs to be recomputed which eventually results into severe computational overheads. This shortcoming is discussed in more detailed next.

## 5.2 Drawback

The volumetric decomposition technique allows for a volumetric representation with significantly reduced complexity. However, the topology of the object is changed as soon as the object is rotated—leading to a different flow volume decomposition. This is a crucial problem, as in most of the real world scenarios, moving objects which frequently change their position are involved. Hence, in such highly dynamic scenarios, the volumetric decomposition needs to be permanently recomputed for every new position, which eventually leads to severe computational overheads. An example illustrates the shortcoming:

*Example 5.3* Figure 5.2a shows an object with a filled cup in the interior before rotating. The graph underneath shows the resulting simple reeb graph. Node 0 represents the volume of the object up to the fill level of the liquid. Node 1 represents the volume above the fill level, while node 2 represents the interior of the cup containing the liquid.

Figure 5.2b shows the object (and the resulting reeb graph) after a 45° rotation. The topology is now completely different, since node 3 represents the volume up to the new filling level of the liquid. The volume of node 5 is significantly larger than the volume of node 1, while node 4 is smaller compared to node 2. Hence, if the objects are rotated, correspondingly adjusted representations have to be created.

This shortcoming is especially crucial in real world scenarios, in which typically transient movements are involved.

*Example 5.4* Consider the application of EPD coatings, where objects like car assemblies or entire car bodies are moved through the tank filled with liquid. The object is dipped into the tank by a certain kinematic, in which various rotation angles are applied throughout the process. In fact, in this industrial application, a typical process consists of a full 360° rotation of the object through the tank. Since empirical evaluations have shown that time steps of maximum 5° rotation give the

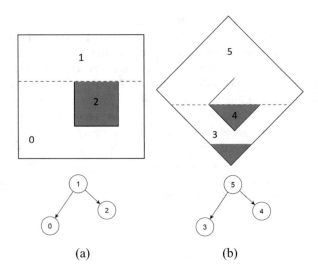

**Fig. 5.2** Influence of rotation to volumes. (**a**) Initial state. (**b**) After rotation

most accurate results, a 360° rotation yields 72 discrete time steps. For each of these time steps a new reeb graph needs to be created.

Obviously, this drawback results in large execution times (especially for complex input data) which poses one of the main problems of volumetric decomposition methods thus far. For that purpose, the following chapter, introduces a parallel scheme which addresses this problem by allowing for concurrent computations of these volumetric decompositions. This allows to overlap the computations of the respective decomposition, and, by that, significantly reduce the computational time.

# Chapter 6
# Volumetric Decomposition on Shared Memory Architectures

As discussed in the previous chapter, the main idea of *volumetric decomposition methods* is to use fewer and larger volumetric units as compared to common volumetric representations used in *Computational Fluid Dynamics* (CFD) methods. However, this volumetric decomposition has to be applied frequently when the object is rotated—leading to severe computational overheads. This constitutes one of the major drawbacks of this method thus far.

To address this shortcoming, a parallel scheme for this volumetric decomposition method is proposed in this chapter (initially introduced in [94]). The proposed scheme introduces two layers of parallelism, which are implemented by employing OpenMP—allowing for the execution on shared memory multi-core architectures. The outer layer enables the parallel construction of the volumetric decomposition, while the inner layer enables parallelism inside the decomposition methodology.

The applicability of this scheme is evaluated in the context of the simulation of air bubbles and liquid drainage in the *Electrophoretic Deposition* (EPD, [6, 61]) process. In EPD, car assemblies or entire car bodies (also known as *Body in White* or BIW for short) are moved through a tank of liquid to apply coatings for corrosion protection. However, issues such as entrapped air bubbles frequently lead to uncoated areas, which is why there is a high demand for an accurate and fast simulation tool.

For that purpose, the proposed scheme is explicitly applied to compute the corresponding volumetric decomposition concurrently for each discrete time step of the simulation. By that, the computations of the volumetric decompositions are overlapped, which significantly reduces the computational time. Experimental evaluations confirm the efficiency and scalability of the proposed scheme, e.g. by showing that the simulation time of a typical use case in the automotive industry could be reduced from almost 6 days to 15 h when employing 16 processing cores.

The remainder of this chapter is structured as follows: The following section provides an overview of the state of the art and the proposed basic architecture. Afterwards, Sect. 6.2 provides a detailed treatment of the proposed parallel simu-

© The Author(s), under exclusive license to Springer Nature Switzerland AG 2021
K. Verma, R. Wille, *High Performance Simulation for Industrial Paint Shop Applications*, https://doi.org/10.1007/978-3-030-71625-7_6

lation scheme. Finally, Sect. 6.3 summarizes the results obtained by experimental evaluations, before this chapter is concluded in Sect. 6.4.

## 6.1  Background

In this section, the background for the simulation of air bubbles and liquid drainage in the process of *Electrophoretic Deposition* (EPD, [6, 61]) which is based on methods for *Computational Fluid Dynamics* (CFD, [2, 82, 98]) is reviewed. In particular the main characteristic of previously proposed methods which, originally, prevented an efficient simulation and motivates the consideration of a so-called volumetric decomposition scheme are reviewed. While this scheme addresses a major obstacle for an efficient simulation, it causes other drawbacks which are discussed afterwards. Resolving this drawback and, by this, eventually enabling an efficient air bubbles and liquid drainage simulation for EPD is then considered in the remainder of this chapter.

### 6.1.1  State of the Art

In EPD processes, objects (e.g. BIWs) get dipped through a tank of paint to apply corrosion protection. However, the emergence of air bubbles while dipping into the tank may prevent a complete coverage of the surfaces. Moreover, entrapped liquid during dipping out may lead to corrosion in one of the consecutive manufacturing processes. In order to simulate the emergence of air bubbles and liquid drainage of such processes, a three-dimensional representation of the object which can serve as input data is necessary. In manufacturing processes, such objects are usually designed using common CAD-tools and, then, exported as meshes which can be used as input for various simulation tools.

Using such a representation, standard tools based on *Computational Fluid Dynamics* (CFD) can be utilized (and, in fact, have been used for many years) in order to simulate corresponding coating processes.

However, CFD is usually applied to simulate a large number of small volumes like meshes composed of tetrahedra or hexahedra [3, 87]. Simulating large objects such as entire car bodies frequently brings CFD to its limits and, hence, typically requires significantly large computation times (even on dedicated HPC clusters). Besides that, CFD is very sensitive to the choice of boundary conditions. A small difference in boundary conditions may lead to a huge deviation in results.

These drawbacks motivate alternative representation of the considered objects which is more suited to the simulation of EPD. For that purpose, the ALSIM architecture has been proposed. This architecture is based on a volumetric decomposition method. In this method, the object gets decomposed into so-called *flow volumes* based on *critical vertices* of the input mesh. As a result, a so-called *reeb graph* accurately represents the topology of the object (as reviewed Sect. 5.1). However, this volumetric decomposition has to be applied frequently when the object is rotated. Empirical evaluations yield that this method is by far the most expensive part of the whole simulation in terms of execution time. In an industrial application, this decomposition typically needs to be applied 72 times[1] resulting in large execution times. Hence, while this scheme addresses a major obstacle for an efficient EPD simulation, it causes other severe drawbacks. Resolving this drawback and, by this, eventually enabling an efficient EPD simulation is then considered in the remainder of this chapter. In the following, the basic architecture of the proposed solution is outlined.

### 6.1.2  Basic Architecture

Generally, the process of simulating air bubbles and liquid drainage in EPD is inherently a sequential process. The simulation consists of a set of discrete time steps $T$, while every discrete time step $0 \leq t < T$ inherently depends on the results of its predecessor. Therefore, the simulation of time step $t - 1$ needs to be completed before the simulation of time step $t$ can be started. This dependency prohibits parallelization in the outermost layer, i.e. independently simulating each time step $t$ in parallel is not possible.

However, the simulation of each time step itself offers potential. In fact, four basic tasks have to be conducted for each time step $t$: (1) primary setup work, (2) creation of the reeb graph, and the actual simulation composed of (3) a hydro-static solving process (rotation) as well as (4) a hydro-dynamic solving process (translation). Table 6.1 summarizes the effort needed for average assemblies for each of these steps with respect to their required execution time.

This distribution of efforts is surprising, considering that EPD is essentially performing fluid simulation, where typically the solving part is rendered as the bottleneck (see e.g. [92, 104]). However, these results clearly suggest that the target of any kind of optimization should be the reeb graph construction. The actual simulation (hydro-static/hydro-dynamic solving) consumes only 15% of the total execution time, which shows that the whole process is clearly dominated by the

**Table 6.1** Distribution of execution times

| Method | Execution time distribution (%) |
| --- | --- |
| Setup | 5 |
| Create reeb graph | 80 |
| Hydro-static solving | 5 |
| Hydro-dynamic solving | 10 |

---

[1]The 72 applications results from the fact that an object is usually rotated the complete 360°, whereby 5° steps are considered sufficient.

volumetric decomposition. Note that for large objects (e.g. BIW), the time spent on setup tasks will be less and shifted towards reeb graph construction.

In order to speedup the process of reeb graph construction, two basic layers of parallelization are introduced:

- Independent parallel computation of each graph: Instead of computing the graph sequentially for every time step $t$ and then directly performing the simulation, $n$ graphs are computed in parallel, while $n$ is the number of cores available on the system. Time step $t$ is then only simulated when the corresponding graph construction is completed and time step $t - 1$ has already been simulated (hereinafter referred to as *outer parallel layer*).
- Parallelization of the reeb graph construction itself: Some of the employed methods for constructing the graph are applicable for data parallelism and are therefore target of a nested parallelism approach (hereinafter referred to as *inner parallel layer*).

The description of the implementation of corresponding techniques is covered by the following section.

## 6.2   Parallel Simulation of Electrophoretic Deposition

This section provides a detailed treatment of parallelization technique. To that end, the *outer parallel layer* and *inner parallel layer* are covered with a detailed description on their respective implementations.

### 6.2.1   Outer Parallel Layer

For the sake of parallelizing the reeb graph construction, the basic flow of the architecture needs to be re-developed. To this end, the reeb graph construction process as applied thus far and sketched in Algorithm 4 is reviewed. The flow consists of mainly three steps while iterating through all time steps $T$. The first step is to rotate the input mesh according to the kinematic of the real process (see Line 3). Based on this rotated mesh, a new reeb graph is created (Line 4). Once this reeb graph is constructed, the hydro-static and hydro-dynamic equation systems are solved (Line 5). Afterwards the results of this time step $t$ are available and can be exported for further analysis (Line 6).

In order to parallelize the reeb graph construction, the base algorithm is re-developed as follows: Instead of iterating through the time steps $t \in T$ and solving the equation system for time step $t$ after creating the graph, $n$ graphs are created in parallel while keeping the equation systems in sequential order. This is sketched in Algorithm 5.

---

**Algorithm 4** Original simulation flow

---

1: $M \leftarrow$ input Mesh
2: **for each** time step $t \in T$ **do**
3:     $M_r \leftarrow rotateMesh(M)$
4:     $G_i \leftarrow createGraph(M_r)$
5:     $G_i \leftarrow solveEquationSystem(G_i)$
6:     exportResults($G_i$)
7: **end for**

---

---

**Algorithm 5** Proposed (parallel) simulation flow

---

1: $V \leftarrow$ empty list
2: **for each** $t \in T$ **do**
3:     $S \leftarrow getStepData(t)$
4:     push $S$ onto $V$
5: **end for**
6: **#pragma omp parallel num_threads($n$)**
7: **#pragma omp parallel for ordered schedule(dynamic, 1)**
8: **for each** $v_i \in V$ **do**
9:     $G_{v_i} \leftarrow createGraph(v_i)$
10:     **#pragma omp ordered**
11:     solveEquationSystem($G_{v_i}$)
12: **end for**

---

Here, the method starts with reading in the step data, which contains the positions and rotation angles of the object (see Line 3). Afterwards, $n$ threads are entrusted with the construction of the reeb graph and the simulation (hydro-static/hydro-dynamic solving) of each time step $t$ (see Line 5–8). As soon as the first graph is completed, the same thread starts with computing the first simulation step (see Line 9–10). Once the second graph has been constructed, the corresponding thread computes the second simulation step and so on. If a thread has completed its simulation step, it will continue with creating more graphs if there are any steps still not simulated. Considering the fact that the graph creation consumes much more computation time than the actual simulation of a step, a thread rarely needs to wait for the previous simulation step to be completed. In fact, this happens only when the graph construction for time step $t$ consumes significantly less time than the simulation of time step $t - 1$.

In general, this method exhibits a rather irregular workload distribution among its iterations. This is because the amount of time needed for the graph construction may differ heavily between the single time steps (depending on the respective rotation angles). Hence, if there are e.g. eight graphs to be constructed and each thread gets assigned two, it might occur that the last graph creation terminates earlier than the first one and, thus, the last thread is just busy waiting. Therefore, to receive optimal performance, a dynamic scheduling paradigm is employed (as shown in Line 6 of Algorithm 5). The actual solving of the equation system needs to be kept in the same order as it was executed in serial, since the result of time step $t$ inherently depends on time step $t - 1$. This behavior is achieved, by employing the ordered clause.

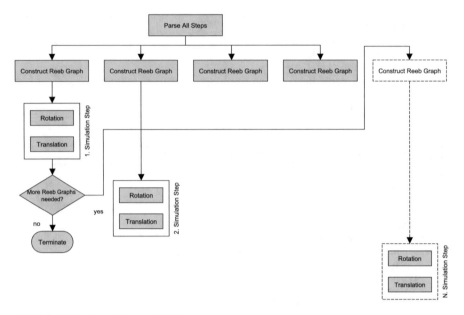

**Fig. 6.1** Illustration of the parallel flow

This still allows for a high degree of concurrency, since the graph construction has substantial run time.

Figure 6.1 illustrates the complete parallel work flow. Each thread gets assigned one reeb graph and, as soon as the corresponding graph construction is completed, time step $t$ is simulated. Once the simulation of time step $t$ terminated, the reeb graph for another time step is constructed if required. Otherwise the thread terminates.

### 6.2.2  Inner Parallel Layer

The second layer focuses on the parallelization of the reeb graph construction itself. To this end, two basic steps have to be considered:

1. Identifying the critical vertices of the mesh (i.e. its local maxima, minima, and saddle points) and,
2. Using this local information, constructing the volume decomposition by a sweep plane algorithm [66].

   Additionally, another step needs to be considered, which was omitted thus far to keep the descriptions simple, namely:
3. Integrating so-called *bottlenecks* into the reeb graph.

   The following descriptions provide details to all three steps.

---

**Algorithm 6** Baseline of collecting extrema values

---

1: $EL \leftarrow$ empty list
2: $M \leftarrow mesh$
3: **for each** vertex $v \in M$ **do**
4:     **if** $v$ is critical **then**
5:         $e \leftarrow$ new extremum induced by $v$
6:         determine region sets and store them in $e$
7:         push $e$ onto $EL$
8:     **end if**
9: **end for**

---

### 6.2.2.1   Identifying Critical Vertices

The first step to construct the reeb graph is to identify the critical vertices, such as local minima, maxima, and saddle points. The basic algorithm iterates through the mesh and creates a new extremum for every identified extrema point. An extrema point can be identified by considering their adjoining points. For example, if adjoining points have lower z-coordinate values, the given point is identified as a local maxima. Algorithm 6 sketches the baseline of the algorithm.

Since the mesh is essentially a set of vertices, this algorithm is parallelized by portioning the mesh container. Each thread iterates through its respective portion and generates a new extremum for every critical point.

### 6.2.2.2   Constructing the Volume Decomposition

As stated above, the volume decomposition is conducted using a plane sweep algorithm. In related work, there have already been several attempts of parallelizing plane sweep algorithms. These attempts traditionally employ methods to statically or dynamically segment the input. The operations are then performed in parallel over these segments. However, some of these proposed solutions introduce the need for heavy synchronization, or can only be applied on input arranged as orthogonal line segments [30, 50]. In [29], the plane sweep methodology is avoided to achieve a solution without the need for synchronization. This approach however does not scale well for a smaller number of cores. In [54], the input is dynamically segmented to spatial operations, while performing the operation on multiple portions of the input in parallel without the need for synchronization. The presented methodology splits the input into strips, while the split lines are generated at roughly equal intervals and are parallel to the sweep plane. Each strip is then used as an input to a sweep plane algorithm.

However, these methods for parallelization are not applicable to the problem considered in this work. Here, the sweep plane algorithm constructs the volume decomposition by scanning for the identified critical vertices. Each of such critical vertex triggers an event, which starts a new flow volume and causes the previous flow volume to end. When separating the input into strips, the information about the previous critical vertex and, hence, the start vertex of a new flow volume is lost. In other words, a parallelization would cause significant data dependencies requiring substantial synchronization efforts and, hence, would basically consume all possible benefits gained by a parallel execution. Consequently, the sweep plane method is kept sequential and no parallelization scheme is suggested for this step.

### 6.2.2.3  Integrating Bottlenecks

The final step of the graph construction is the integration of so-called bottlenecks. These are needed to consider liquid flows with respect to time. Without the consideration of time, liquids would touch every reachable surface of the object immediately. But of course, liquids need time to spread, especially when it goes through narrow channels. For that purpose, it is essential to detect all narrow channels of an object in order to limit the liquid flow speed. Such narrow channels are referred to as *bottlenecks*.

For the purpose of detecting bottlenecks, which are defined as the shortest loop round a narrow channel, an algorithm can be employed which creates a distance field (using the Dijkstra algorithm) in order to find the shortest path between two nodes in a graph. Starting from one vertex of the mesh, its neighbors are added to the distance field, along with their neighbors, until the given circumference is reached. A bottleneck is found whenever the neighborhood of a newly inserted vertex fulfills a certain criterion.

As discussed, the flow volumes and their relations are represented by the reeb graph. Bottlenecks are found and represented as lists of vertices. For the simulation, their position with respect to the flow volumes is relevant. Therefore, the bottlenecks need to get integrated into the reeb graph.

There exist basically two types of bottlenecks, horizontal and vertical bottlenecks, as illustrated in Fig. 6.2a (bottlenecks are depicted as small ellipses). Vertical bottlenecks can only appear inside of a volume, therefore a split of the volume is indispensable. Horizontal bottlenecks can either appear between two volumes or inside a volume. In case they appear inside a volume, the volume needs to be split, otherwise the respective node connection just needs to be marked as a bottleneck connection. This bottleneck integration significantly influences the flow volumes, as seen in Fig. 6.2b. The influence to the reeb graph is shown in Fig. 6.3. Figure 6.3a shows the reeb graph before bottleneck integration. The nodes with dashed lines correspond to volumes which need to be split. Figure 6.3b shows the reeb graph after splitting. The dashed lines show the newly introduced bottleneck connections, through which the liquid flow is limited.

This whole process of bottleneck integration is especially important, considering that empirical evaluations have shown that for a BIW, typically more than 3000 of such bottlenecks are detected. This causes the amount of flow volumes to increase heavily and takes up to 50% of the whole reeb graph generation run time.

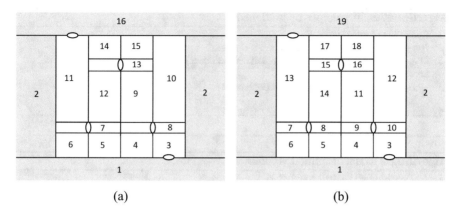

**Fig. 6.2** Integration of bottlenecks into the volumetric decomposition. (**a**) Initial decomposition. (**b**) Decomposition after bottleneck integration

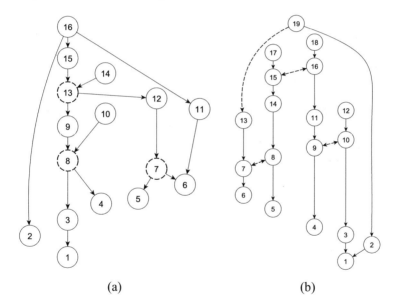

**Fig. 6.3** Influence of bottleneck integration to the reeb graph. (**a**) Initial reeb graph. (**b**) Reeb graph after bottleneck integration

For the purpose of parallelizing this process, the bottleneck integration is split into two basic methods: The volume splitting and the volume connecting. Algorithm 7 sketches the basic algorithm which can be executed in parallel. A map $M$ of volumes $V$ to be split by their associated bottlenecks $B$ is stored (see Line 1–3) . While iterating through the map, the volume $v_i$ is split by its associated bottleneck $b$ (Line 5). The resulting volumes are stored in the list $L_{v_s}$, which is then integrated

---

**Algorithm 7** Split volumes by bottlenecks

---
1: $L_V \leftarrow$ list of all volumes $V$
2: $B \leftarrow$ list of all detected bottlenecks
3: $M \leftarrow$ map of all volumes $V$ with associated bottlenecks $B$
4: **for each** element $e \in M$ **do**
5:     $L_{v_s} \leftarrow$ split $v$ of $e$ by associated bottleneck $b$
6:     integrate $v_s$ into $L_V$
7: **end for**

---

---

**Algorithm 8** Connect splitted volumes

---
1: $L_{V_s} \leftarrow$ list of all splitted volumes $V_s$
2: **for each** splitted volume $v_s \in L_{V_s}$ **do**
3:     connect $v_s$ to upper volumes
4:     connect $v_s$ to lower volumes
5: **end for**

---

into the volume list $L_V$ (Line 6). Since $L_V$ is a global container, which stores all volumes sorted by ascending z-levels, the write operation to this container needs to be sequential and, hence, locked.

After the volumes are split, they need to be connected to their respective upper and lower volumes. This is sketched in Algorithm 8. While iterating through all splitted volumes $L_{V_s}$ (see Line 1–2), a given splitted volume $v_s$ is first connected to the upper volumes and then to its lower volumes (Line 3–4). Since it might happen that the upper or lower volumes of $v_s$ were also splitted, the actual setting of the connection needs to be locked in order to avoid race conditions.

Both algorithms can be executed in parallel and, by this, significantly speed up the process. Evaluations summarized in the next section, confirm this improvement.


## 6.3   Experimental Evaluations

In order to evaluate the performance of the proposed parallel scheme, a range of experiments was conducted whose results are summarized in this section. The following subsection shows the speedup obtained for the reeb graph construction alone. Afterwards, the speedup obtained for the entire simulation is presented.


### 6.3.1   Speedup for the Reeb Graph Construction

To evaluate the scalability of the methods in terms of input size and number of execution cores, data sets composed of different numbers of triangles were executed

**Table 6.2** Considered data sets

|  | Spare wheel case | Liftgate | Cabin | BIW |
|---|---|---|---|---|
| # triangles | 60k | 200k | 850k | 3M |

employing a range of up to 16 cores. Table 6.2 shows the data sets considered for the experiments. The number of triangles refers to the size of the input triangular surface mesh. The considered data sets are typical data sets used in the automotive industry: *Spare wheel case*, *liftgate*, and *cabin* are car assemblies (i.e. car parts), while BIW (*Body In White*) refers to an entire car body.

All experiments have been executed on a two socket Intel Xeon E5-2660 v3. The source code was compiled with GCC 5.4.0 with optimization level -O3 and executed on Ubuntu 16.04. Additionally, a *fill-socket-first* policy was adopted. Both, the *outer parallel layer* and the additional *inner parallel layer* (as introduced in Sects. 6.2.1 and 6.2.2, respectively) are compared with respect to speedup.

Figure 6.4a shows the obtained speedup without including the *inner parallel layer*. The values show that for all considered test cases, this basic parallel approach results in a significant speedup. However, for smaller data sets the efficiency drops exceedingly when increasing the number of cores (e.g. for the *spare wheel case* 16 cores yield the same speedup as 8 cores). This is caused by the fact that, for smaller data sets, the time spent on reeb graph construction is proportionally smaller compared to bigger data sets and shifts more towards the actual simulation. Hence, some threads are just busy waiting after completing their respective reeb graph construction until the simulation of the preceding step is completed. For the entire car body (BIW), a speedup of 12.3 was achieved when employing 16 cores.

Figure 6.4b shows the speedup obtained when additionally including the *inner parallel layer*. The values show that in almost all the considered test cases, this approach yields an additional speedup compared to using only the basic parallel scheme. Only for the smallest data set, the *spare wheel case*, this approach does not excel, particularly when employing higher number of cores. For the largest data set, the BIW, a speedup of 14.6 was achieved when using 16 cores.

The presented results show that the introduced methods yield significant speedups for all ranges of input sizes when employing a smaller number of cores. For bigger data sets, also a higher number of cores scales well. However, a higher number of cores also introduces an overhead for smaller data sets resulting in less efficiency for the considered smaller test cases.

## 6.3.2 Speedup for the Entire Simulation

To show the improvements gained for a typical industrial automotive use case, the speedup in absolute times of the entire simulation (i.e. all steps listed in Table 6.1) gained for a BIW is shown in addition. The BIW is composed of 2.5 million triangles, the simulation is using 72 discrete time steps of 5° rotation each.

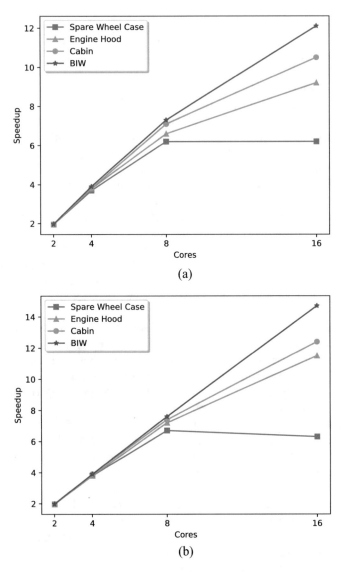

**Fig. 6.4** Speedup of the parallel reeb graph construction scheme. (**a**) Without including the *inner parallel layer*. (**b**) With including *inner parallel layer*

Figure 6.5 shows the timings in hours received for the BIW, including both, the *inner parallel layer* and the *outer parallel layer*. While a sequential simulation consumes almost 6 days, the parallel simulation employing 16 cores can be conducted within 15 h.

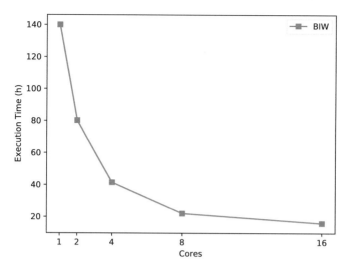

**Fig. 6.5** Speedup of the entire simulation in absolute execution time

## 6.4    Summary

In this chapter, a parallel scheme for a volumetric decomposition method for shared memory architectures has been presented. Starting from a sequential architecture, potential parallelism was unveiled in the process of reeb graph construction. In order to speedup this construction, two parallelization layers where introduced which were implemented in C++ employing OpenMP. The presented methods yield significant speedups for the reeb graph construction for both, smaller as well as larger data sets. For an entire car body, a speedup of 14.6 was achieved for the reeb graph construction when employing 16 cores. The execution time of the entire simulation could be reduced from almost 6 days to 15 h for a typical automotive industry use case where a BIW is considered.

# Chapter 7
# Volumetric Decomposition on Distributed Memory Architectures

As reviewed in Chap. 5, the volumetric decomposition method allows for a volumetric representation with significantly reduced complexity as compared to standard *Computational Fluid Dynamics* (CFD) methods. However, the topology of the object is changed as soon as the object is rotated—leading to different volumetric decomposition. This is a crucial problem, as in most real world scenarios, moving objects that are constantly rotated are involved. Eventually, this leads to severe computational overheads.

To overcome this drawback, in the previous chapter, a parallel framework on a threading level has been presented. This framework allows to compute the volumetric decomposition of each discrete time step in parallel. Following this flow, the computational time is reduced significantly. However, this approach still suffers from two aspects: (1) high memory consumption as a result of concurrently storing multiple reeb graphs in the memory, and (2) limited parallel computing cores due to hardware limitations of shared memory architectures. To make this parallel approach practically more applicable, the execution on distributed memory architectures needs to be enabled.

In this chapter, the parallel framework is extended for distributed parallel architectures based on the MPI framework [27, 31] (originally proposed in [95]). In order to accomplish that, the following major aspects in this regard are considered: (1) The initial workload distribution of the discrete input rotation degrees, (2) the inherent serial dependency of the concrete time step simulation, (3) the high memory requirement of the employed graph structure, and (4) the irregular workload distributions.

The correspondingly obtained method is described in the remainder of this chapter as follows: First, Sect. 7.1 briefly introduces the proposed distributed memory architecture for the volumetric decomposition method. Afterwards, Sect. 7.2 describes how the parallel scheme is improved by addressing the four aspects listed above—eventually leading to an efficient simulation of air bubbles and liquid drainage in EPD on distributed memory architectures. Finally, Sect. 7.3 summarizes

---

**Algorithm 9** Basic simulation flow

---

1: $M \leftarrow$ input Mesh
2: **for each** time step $t \in T$ **do**
3:　　$M_r \leftarrow rotateMesh(M)$
4:　　$G_t \leftarrow createGraph(M_r)$
5:　　$G_t \leftarrow simulate(G_{t-1}, G_t)$
6:　　exportResults($G_t$)
7: **end for**

---

the results obtained from experimental evaluations, before the chapter is concluded in Sect. 7.4.

## 7.1　Basic Architecture

As introduced in the previous chapter, the simulation of air bubbles and liquid drainage in EPD follows an inherent serial flow. The corresponding basic serial simulation flow employed thus far is sketched in Algorithm 9. This flow consists of mainly three steps while iterating through all time steps $T$. The first step is to rotate the input mesh according to the kinematic of the real process (see Line 3). Based on this rotated mesh, a new reeb graph is created (Line 4). Once this reeb graph is constructed, the actual simulation (hydro-static and hydro-dynamic solving) is conducted (Line 5). Afterwards the results of this time step $t$ are available and can be exported for further analysis (Line 6).

In order to allow the execution of the EPD air bubble and liquid drainage simulation on distributed parallel architectures, this basic simulation flow is re-developed. In the distributed setup, each process is performing the computations of the time steps $t \in T$ independently. For that purpose, each process initially requires two inputs: (1) the original input mesh and (2) the input time step $t \in T$ to be computed. Therefore, at first, the input time steps are subdivided and distributed to the independent processes. This data contains the rotation angles of the input object of each discrete time step. The rotation will be performed on the original input mesh, which is also distributed to the individual processes.

However, as already discussed, the simulation of time step $t - 1$ needs to be completed before the simulation of time step $t$ can be started. Therefore, each process not only requires the original input data, but also the result of the preceding simulation step. Based on this preceding result, the current process can compute its corresponding simulation step.

Overall, for each process this yields the basic workflow as sketched in Algorithm 10. Here, each process is assigned a partition of the input time steps $t \in T$, which contains the positions and rotation angles of the object (see Line 1). Based on this data, each process independently rotates the input mesh and constructs the corresponding reeb graph (Line 4–5). In order to simulate this time step, the process needs to receive the preceding simulation result, based on which the current step can

---

**Algorithm 10** Proposed distributed simulation flow

---

1: $V \leftarrow$ assigned time steps
2: $M \leftarrow$ input Mesh
3: **for each** $v \in V$ **do**
4:     $M_r \leftarrow rotateMesh(M)$
5:     $G_v \leftarrow createGraph(v)$
6:     **recv** $G_{v-1}$
7:     $G_v \leftarrow simulate(G_{v-1}, G_v)$
8:     **send** $G_v$
9:     exportResults($G_v$)
10: **end for**

---

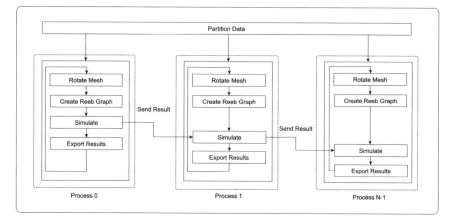

**Fig. 7.1**   Illustration of the parallel flow

be simulated (Line 6–7). The results are stored inside the reeb graph and dispatched to the subsequent process, as well as exported to disc for further analysis (Line 8–9). This basic workflow is also illustrated in Fig. 7.1.

However, to receive optimal performance, it is key to consider certain important aspects of the implementation. More precisely, at first, the initial workload distribution of the discrete input rotation degrees needs to be considered. Moreover, the inherent serial dependency of the concrete time step simulation and the high memory requirement of the employed graph structure further increase the complexity. Finally, the irregular workload distributions among the concrete time steps need to be taken into account to receive optimal performance. The details for this are covered next.

## 7.2   Implementation of the Distributed Algorithm

In order to efficiently implement the proposed workflow, the four aspects as described above need to be considered. To this end, the following section covers the implementation of the workload distribution, as well as a dedicated memory optimization strategy. Moreover, load balancing issues and limitations of the presented scheme are discussed.

## 7.2.1 Workload Distribution

The general setup of the distributed algorithm is based on a master-slave architecture. Instead of iterating through the time steps $t \in T$ and solving the equation system for time step $t$ after creating the graph (as sketched in Algorithm 9), the first process ($N_0$) starts by partitioning the input data and distributing it to the remaining processes. Due to the inherent dependency of time step $t$ on $t-1$, an optimal partition should allow a contiguous computation of the time steps $t \in T$. Therefore, time steps $t_0$ to $t_{p-1}$ should be the first steps to be computed, where $p$ is the total number of processes. Hence, the data for each process $N$ is partitioned as

$$N_0 = \{t_0, t_{0+p}, t_{0+2p}, \ldots, t_{0+(T-p)}\},$$
$$N_1 = \{t_1, t_{1+p}, t_{1+2p}, \ldots, t_{1+(T-p)}\},$$
$$N_{p-1} = \{t_{p-1}, t_{p-1+p}, t_{p-1+2p}, \ldots, t_{p-1+(T-p)}\},$$

where $p$ is the total number of processes and $T$ the total number of time steps.

However, in some certain kinematics it might occur that identical rotation degrees are used within one simulation. For example: (1) when the object dips in, an identical rotation degree might occur during the dip out phase, or (2), when a specific rotation degree needs special consideration and is therefore splitted into small time slices to receive more accurate results. In such cases, the reeb graph from the duplicated rotation degree could be re-used and does not need to be constructed again. For that purpose, the identical input rotation angles are assigned to the same process, such that the reeb graph for the duplicated rotation degree does not need to be created again, but can be re-used instead.

*Example 7.1* Figure 7.2 shows input data containing rotation degrees for 10 time steps which should exemplarily be partitioned to three processes. Each of the colored fields marks one tuple with identical rotation degrees. During the iteration through the input, the individual rotation degrees are assigned consecutively to the processes. The duplicated rotation degrees are assigned to the same process to allow re-usage of the constructed reeb graph.

After this initial data partitioning, the first process sends: (1) the original input mesh, and (2) the data partition to the corresponding processes. Asynchronously to this memory transfer, the first process starts the creation of the reeb graph of time step $t_0$ and the subsequent simulation. Once the remaining processes have received their corresponding input data, each process is computing its rotated mesh, based on which the reeb graph is constructed. As soon as the simulation of time step $t_0$

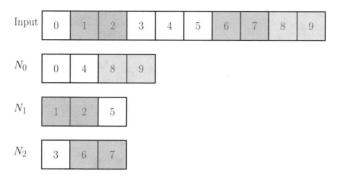

**Fig. 7.2** Partition of the input data to three processes. Colored fields mark identical rotation values

is completed by the first process $N_0$, its result is transferred to the subsequent process $N_1$. Based on the simulation result of $t_0$, process $N_1$ can conduct its corresponding simulation of time step $t_1$ and once completed, dispatch the result to process $N_2$. This flow is repeated until all time steps $t \in T$ are simulated.

By employing this communication flow, the actual simulation of the time steps $t \subset T$ are kept in the same order as they were executed in serial.

However, not only the inherent serial dependencies, but also the high memory consumption of the reeb graph renders the dispatching of the simulation result to the consecutive processes a non-trivial task. To receive optimal performance, further optimization on the memory consumption is required, as discussed next.

## 7.2.2 Memory Optimization

The high memory requirement of the reeb graph method yields a major drawback in a distributed setup, where the reeb graph needs to be dispatched between the processes. Although the reeb graph method allows for efficient volumetric representation with reduced complexity as compared to volumetric mesh approaches, it suffers from high memory requirements as summarized in Table 7.1. Here, the memory requirements for typical data sets used in the automotive industry: *Spare wheel case*, *liftgate*, and *cabin* are car assemblies (i.e. car parts), while BIW (*Body In White*) refers to an entire car body. Considering that in a parallel approach using $n$ processes, $n$ reeb graphs will be allocated at the same time, this clearly suggests that the reeb graph construction needs to be distributed among a dedicated cluster. In this distributed setup, however, the reeb graphs frequently need to be dispatched between the processes, since the simulation result is incorporated into the graph.

Figure 7.3 shows a simple object with the corresponding reeb graph in two discrete time steps with different rotation angles. The simulation result is incorporated into the reeb graph in a way that each reeb graph nodes stores its corresponding liquid filling level. Hence, in Fig. 7.3a reeb graph node 2 is fully filled, while 0

**Table 7.1** Reeb graph
memory requirements

| Data set | # Triangles | Memory requirement |
|---|---|---|
| Spare wheel case | 60k | 150 MB |
| Liftgate | 200k | 450 MB |
| Cabin | 850k | 1.3 GB |
| BIW | 3M | 6 GB |

**Fig. 7.3** Influence of rotation
to the reeb graph. (**a**) Initial
state. (**b**) After rotation

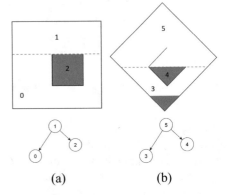

(a)                    (b)

and 1 are completely empty. In Fig. 7.3b, reeb graph node 3 and 4 are partially
filled and node 5 is completely empty. Based on this reeb graph, the simulation
result is represented and can also be projected back onto the original input mesh.
Therefore, the reeb graph is also employed to transfer the result from one simulation
step to another. In Fig. 7.3 it is also shown that the reeb graph is heavily influenced
by the corresponding rotation degree. The volumetric decomposition of Fig. 7.3b
significantly differs from the decomposition before rotation as shown in Fig. 7.3a.
Due to that, the transfer of simulation results from one reeb graph to another is a
non-trivial task. In order to transfer the result, the connection from one reeb graph
node to the corresponding reeb graph node after rotation needs to be computed.
However, as shown in Table 7.1, the reeb graph data structure is extremely heavy in
terms of memory requirement. Since dispatching such heavy data structure among
a distributed architecture is highly inefficient, a simplified reeb graph has been
developed which stores the following data:

1. In order to compute the relation between two rotated reeb graphs, the vertices
   of the original input mesh are employed. Every reeb graph node represents one
   volume of the underlying mesh. To compute the relation after rotation, the IDs
   of the vertices of the underlying mesh are stored.
2. For every vertex of the underlying mesh, every reeb graph node additionally
   stores a Boolean which represents if the corresponding vertex is in touch with
   liquid or not.
3. For each reeb graph node, the corresponding liquid filling level is stored.

This simplified reeb graph representation has significantly less memory requirement than the standard reeb graph, while the preceding simulation result can still be accurately represented.

While this reduced memory requirement essentially simplifies the communication among the distributed memory architecture, the method still exhibits a rather irregular workload distribution among their iterations caused by the inherent serial dependency. The details on this are discussed next.

## 7.2.3 Load Balancing

The whole simulation can essentially be abstracted by three major tasks: (1) the reeb graph construction, (2) the actual simulation (hydro-static and hydro-dynamic equation systems), and (3) various setup and I/O tasks. Out of these three major tasks only the reeb graph construction offers potential parallelism, while the other tasks are inherently serial. The ratio between the reeb graph construction and the other tasks, hence the ratio between parallel and serial tasks is roughly 80% to 20%. Figure 7.4 illustrates an abstract task view of the whole workflow. The serial part has been summarized by the *Simulation* task. The illustration shows how this inherent serial dependency of time step $t$ to $t - 1$ strongly limits the parallel scalability of the whole process. The higher task numbers yield a large idle time between the reeb

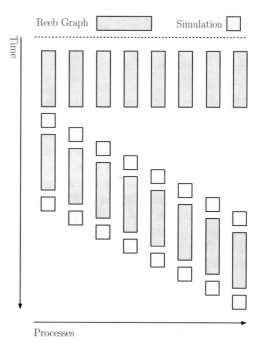

**Fig. 7.4** Abstract processes view of the whole parallel scheme

Reeb Graph          Simulation

Time

Processes

graph construction and the actual simulation. This strongly suggest that this scheme cannot scale well on a higher number of processing cores. Furthermore, it is also evident that further optimization to the reeb graph construction itself will not result in much higher speedup. Regardless of the magnitude of improvement, the limiting factor is the inherent serial work, as described by Amdahl's Law.

Since this inherent serial dependency cannot be overcome, the only way to reduce this idle time would be to reduce the time of the serial work, hence the actual simulation time. However, in this process a hydro-static and hydro-dynamic equation system is solved which itself offers very limited potential to parallelism. Therefore, the presented scheme of parallelizing the reeb graph construction on a distributed memory architecture is close-to-optimal considering the inherent limiting factors of the application. Evaluations summarized in the next section, confirm these enhancements.

## 7.3  Experimental Evaluations

In order to evaluate the performance of the proposed methods, a range of experiments have been conducted whose results are summarized in this section. More precisely, the obtained speedup is presented for the reeb graph construction alone (i.e. excluding the actual simulation) as well as for the entire simulation process. Before results on that are presented, however, the utilized test environment and the considered data sets are described first.

### 7.3.1  Test Environment and Considered Data Set

The experiments have been conducted on a cluster containing eight individual nodes. Each of the nodes contains two Intel Xeon E5-2620 v4 2.10 GHz with eight physical cores each. The source code was compiled with GCC 5.4.0 with optimization level -O3 and executed on CentOs 7.

To evaluate the scalability of the methods with respect to the input size and number of processes, data sets of different sizes, i.e. composed of different numbers of triangles forming the surface mesh, have been considered. All of them constitute typical data sets used in the automotive industry such as a *Spare wheel case*, a *Liftgate* and a *Cabin* (which all represent parts of a car) as well as an BIW (i.e. a *Body In White*) which represents an entire car body (and, hence, is the largest data set). Table 7.2 provides the number of triangles for each of those data sets.

**Table 7.2**  Considered data sets

|             | Spare wheel case | Liftgate | Cabin | BIW  |
|-------------|------------------|----------|-------|------|
| # triangles | 60k              | 200k     | 850k  | 3.4M |

Each data set is considered for 72 discrete time steps which yields a simulation of a complete rotation of 360° assuming that simulation steps after each 5° are considered sufficient. For each of the data sets, the results of the proposed distributed method are compared with the results obtained by the serial method, i.e. using a single processing core.

## 7.3.2 Speedup in the Reeb Graph Construction

Table 7.3 shows the speedup and efficiency obtained for the four considered data sets using the proposed distributed method compared to the serial method.

The results are also visualized in Fig. 7.5. The obtained values show that the best speedup is achieved when using 32 processes. Then, e.g. for the Cabin, a speedup of 13.1 and, for the BIW, a speedup of 16.2 is obtained. Only for the smallest data set, the spare wheel case, 16 processes yield the best speedup of 6.1. Generally, for the spare wheel case the obtained speedup is significantly lower as compared to large data sets. This results from the fact that for smaller data sets, the execution time for the reeb graph construction is proportionally smaller as compared to the time spent on synchronization efforts. This generically renders the obtained speedup for smaller data sets under-performing as compared to larger data sets. For the larger data sets it is shown that an efficiency of over 0.9 is obtained when employing up to eight processes. For the largest data set, the BIW, even for 16 processes an efficiency of 0.93 is achieved.

**Table 7.3** Results obtained for the reeb graph construction

(a) Speedup

| Processes | Spare wheel case | Liftgate | Cabin | BIW |
|---|---|---|---|---|
| 2 | 1.9 | 1.9 | 1.9 | 1.9 |
| 4 | 3.7 | 3.8 | 3.8 | 3.9 |
| 8 | 5.7 | 7.5 | 7.6 | 7.8 |
| 16 | 6.1 | 12.2 | 12.9 | 14.8 |
| 32 | 5.7 | 12.1 | 13.1 | 16.2 |
| 64 | 3.5 | 4.7 | 5.2 | 7.3 |

(b) Efficiency

| Processes | Spare wheel case | Liftgate | Cabin | BIW |
|---|---|---|---|---|
| 2 | 0.98 | 0.98 | 0.99 | 0.99 |
| 4 | 0.93 | 0.95 | 0.95 | 0.98 |
| 8 | 0.71 | 0.94 | 0.95 | 0.98 |
| 16 | 0.38 | 0.76 | 0.81 | 0.93 |
| 32 | 0.18 | 0.38 | 0.41 | 0.51 |
| 64 | 0.05 | 0.07 | 0.08 | 0.11 |

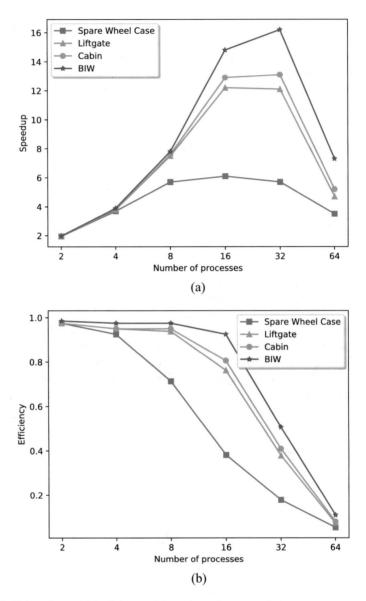

**Fig. 7.5** (a) Speedup and (b) efficiency of the reeb graph construction

However, the obtained values also strongly suggest that, for more than 32 processes, the speedup drops drastically. This originates from the fact that in total 72 discrete time steps are considered in the experiments and, hence, 72 reeb graphs need to be computed. When employing 64 processes, 64 reeb graphs will be computed in parallel. After those are completed, only eight reeb graphs are missing, yielding workload for eight processes, whereas the remaining processes remain idle. This essentially results in dramatic drop of efficiency, e.g. 0.11 for the BIW when employing 64 processes.

### 7.3.3 Speedup in the Entire Simulation

To show the improvements gained for a typical industrial automotive use case, the speedup in absolute times of the entire simulation gained for a BIW is presented. The BIW is composed of three million triangles, the simulation is considering 72 discrete time steps of $5°$ rotation each.

Table 7.4 shows the obtained values for absolute execution time hours as well as speedup and efficiency compared to the serial method. The results are also visualized in Fig. 7.6. The results confirm that the proposed method yields significant improvements. While the serial simulation took 192.5 h, the simulation method proposed in this work utilizing 32 processes terminated in 14.9 h, resulting in a speedup of 12.9. When 64 processes are used, the presented approach does not excel due to the inherent serial dependencies of the application. In a simulation using 72 time steps, the 64 processes are frequently just busy waiting after completing their respective reeb graph construction until the simulation of the preceding step is completed. Therefore, the efficiency for 64 processes is merely 0.08, as presented in Table 7.4.

However, for up to 32 processes, significant speedups with satisfying efficiency is obtained. In fact, the time needed to simulate the industrial example considered here can be reduced from over 8 days to just less than 15 h.

**Table 7.4** Results obtained for the entire simulation

| Processes | Absolute (h) | Speedup | Efficiency |
|---|---|---|---|
| 1 | 192.5 | 1 | 1 |
| 2 | 106.9 | 1.8 | 0.9 |
| 4 | 58.3 | 3.3 | 0.82 |
| 8 | 29.6 | 6.5 | 0.81 |
| 16 | 15.4 | 12.5 | 0.78 |
| 32 | 14.9 | 12.9 | 0.40 |
| 64 | 35.6 | 5.4 | 0.08 |

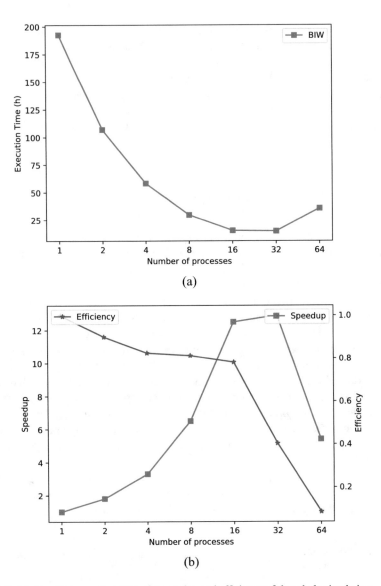

**Fig. 7.6** (**a**) Absolute execution time, (**b**) speedup and efficiency of the whole simulation

## 7.4   Summary

In this chapter a distributed parallel scheme for an industrial electrophoretic deposition simulation has been presented. Dedicated workload distribution and memory optimization methods were presented and implemented in C++ employing the MPI framework. The proposed methods result in significant speedup for the entire simulation. For an industrial use case simulation using a full car body (Body In White), a serial simulation consumed over 190 h, whereas the parallel approach could be completed in less than 15 h using 32 processes. Future work includes the extension of the simulation with new physical phenomena, as well as geometrical optimizations for the reeb graph construction.

# Part IV
# Particle-Based Methods

This part of the book covers simulation methods based on the *Smoothed Particle Hydrodynamics* (SPH). As reviewed in Sect. 2.1, in SPH the fluid is discretized by sample points, which are identified as particles. These particles completely define the fluid, which means that the particles move with the fluid. The particles carry certain physical quantities which are computed based on weighted contributions from neighboring particles. There exist several variants of SPH, each of which yielding different advantages and disadvantages in certain areas of application. Generally, SPH yields strong capabilities in handling multi-phase flows or large deformations. Since, in SPH, the domain is modeled by particles which move naturally according to physical laws, there is no need to track any interface (i.e. the layer between two different fluid phases). Moreover, the tedious mesh generation can be omitted, as the whole domain is completely defined by particles.

However, despite these benefits, SPH suffers from a huge numerical complexity. In order to simulate real life phenomena in appropriate resolution, up to several hundred millions of particles are necessary. Thus, simulations of short periods of physical time frequently require large execution times. At the same time, the discrete particle formulation makes SPH applicable for *High Performance Computing* (HPC) methods, which eventually allows to apply the SPH technique to more practically relevant and industrial applications.

In this context, the use of methods for *General Purpose Computation on Graphics Processing Units* (GPGPU) is particularly suited. These methods yield the advantage that they allow to fully exploit the computational power of GPUs. While CPUs are typically composed of just a few cores, GPUs are composed of thousands of processing cores. Hence, computations on GPUs can be executed on thousands of cores simultaneously. Since the discrete particle formulation of SPH allows for independent computations in a massively parallel manner, the use of GPGPU methods results in significant performance improvements as compared to classical CPU computations.

However, for large-scale SPH simulations in which millions of particles need to be considered, multiple GPU devices need to be utilized, yielding dedicated multi-GPU architectures. Here, a spatial subdivision is employed that partitions the whole simulation domain into individual subdomains which are then distributed among the GPUs and executed in parallel. Since, as discussed above, in any SPH simulation neighboring particles need to be considered, a dedicated communication strategy between the subdomains is required. This exchange of particle information between neighboring subdomains is a huge communication burden in any multi-GPU architecture and rendered as one of the major bottlenecks. Moreover, the inherent moving nature of the particles leads to the fact that particles move between subdomains, which frequently results in unbalanced workloads. Overall, these severe communication burdens and unbalanced workloads result in overheads that, thus far, prevents to fully exploit the potential of HPC in general and GPGPU in particular for SPH applications.

To overcome these shortcomings, this part of the book presents advanced GPGPU computing methods for large-scale industrial SPH simulations. To that end, a dedicated multi-GPU architecture is developed that aims to overcome the inherent communication burdens of such massively parallel architectures. For that purpose, a load balancing methodology with minimal overhead is presented. This load balancing methodology dynamically balances workload between subdomains even on heterogeneous multi-GPU architectures consisting of different types of GPUs. This eventually allows to gain significant performance improvements compared to straightforward approaches.

Moreover, different variants of the SPH technique are considered, in particular, the so-called *Predictive-Corrective Incompressible SPH* (PCISPH) method. PCISPH uses a dedicated predictive-corrective scheme to predict certain physical quantities, which eventually allows to outperform other SPH variants by up to one order of magnitude. However, this predictive-corrective scheme results in increased synchronization effort, which eventually yields severe shortcomings employing a straightforward parallelization scheme. Due to this shortcoming, PCISPH did not find applicability to model real world phenomena thus far.

In order to avoid hat, this part also presents a PCISPH implementation that is suitable for multi-GPU architectures and eventually allows for PCISPH to being applied to real world scenarios. To that end, a revised PCISPH workflow is introduced, which facilitates large hardware utilization and minimal idle time of the underlying GPUs. Moreover, dedicated optimization techniques are considered which lead to significantly increased performance and to eventually fully exploit the inherent advantages of PCISH.

To set the context for the contributions provided in this part of the book, Chap. 8 provides the basics of SPH simulations, including the fundamental formulations, as well as the resulting challenges. Then, Chap. 9 presents the developed multi-GPU techniques aiming to overcome these challenges. Finally, Chap. 10 introduces a novel dedicated multi-GPU architecture for a specific SPH variant.

# Chapter 8
# Overview

## 8.1 SPH Fundamentals

*Smoothed Particle Hydrodynamics* (SPH) is a particle-based, fully Lagrangian method for fluid-flow modeling and simulation. Nowadays, the SPH approach is increasingly used for simulating hydro-engineering applications—involving free-surface flows where the natural treatment of evolving interfaces makes it an enticing approach.

In this section, the basics of the SPH method are introduced. This includes the basic SPH formulations as well as the modeling of the respective internal and external forces.

### 8.1.1 *Formulation*

The SPH method was independently proposed by Gingold and Monaghan [28] to simulate astrophysical phenomena at the hydrodynamic level (compressible flow). The main ideas of the SPH rely on the following basis: Let $J$ be the set of all considered discrete particles. Then, a scalar quantity $A$ is interpolated at position $r$ by a weighted sum of contributions from $J$, i.e.

$$\langle A(r)\rangle = \sum\nolimits_{j\in J} A_j V_j W(r - r_j, 2h), \tag{8.1}$$

where $V_j$ is the volume of the respective particle $j$, $r_j$ is the position of this particle, and $A_j$ the field quantity at position $r_j$. $W$ is a smoothing kernel with the so-called smoothing length $2h$ as a width—defining that only particles within a distance shorter than $2h$ will interact with a particle $j$. This kernel function $W$ is a central part of SPH simulations and the appropriate choice of a smoothing kernel for a specific problem is of great importance. At the same time, a kernel must satisfy three conditions/properties, namely

1. the *normalization condition*

$$\int_r W(r - r_j, 2h)dr = 1 \qquad (8.2)$$

stating that the integral over its full domain is unity,
2. the *Delta function property*

$$\lim_{h \to 0} W(r - r_j, 2h) = \delta(r - r_j) \qquad (8.3)$$

stating that, if the smoothing length $2h$ approaches zero, a delta distribution is applied (with $\delta$ being the Dirac delta function), and
3. the *compact support condition*

$$W = 0 \text{ when } |r - r_j| > 2h \qquad (8.4)$$

ensuring that only particles within the smoothing length $2h$ are considered.

*Example 8.1* Figure 8.1 illustrates a 2D domain with a kernel function $W$ and smoothing length $2h$ for a particle $j \in J$.

Such a Gaussian kernel function as illustrated yields straightforward mathematical properties, however, may not always be the optimal kernel to be employed. For a more detailed treatment on that, the reader is referred to [53, 58].

In Eq. (8.1) it is shown that any continuous quantity is interpolated at a position $r$ by a weighted sum of contributions. Here, it is assumed that the particle masses

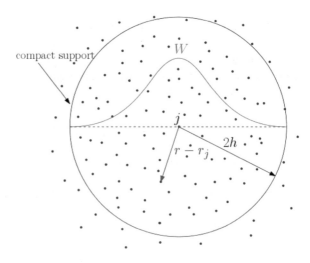

**Fig. 8.1** Illustration of a 2D domain for a particle $j \in J$

$m_j$ for all particles are known beforehand. However, the particle density $\rho_j$ depends on the particle mass $m_j$ and must be computed based on the continuous field of the fluid. By employing Eq. (8.1), the density consequently yields

$$\rho_j = \sum_{j \in J} m_j W(r - r_j, 2h), \tag{8.5}$$

where $\rho_j$ is the particle density of the respective particle $j \in J$, $m_j$ is the particle mass.

## 8.1.2 Internal Forces

Forces can be categorized into internal and external forces. Internal forces are force contributions that occur within the fluid (e.g. pressure or viscosity force), whereas external forces occur from outside the fluid (e.g. gravity of buoyancy force). These external and internal forces are balanced against each other and can be combined into a sum of force terms.

The quantities of the corresponding forces are computed by applying the SPH formulations.

To that end, the internal pressure force $\mathbf{f}_j^{pressure}$ can be obtained by the pressure $p$ at a particle $j \in J$. Based on Eq. (8.1) the pressure force thus yields

$$\mathbf{f}_j^{pressure} = -\sum_{j \in J} p_j \frac{m_j}{\rho_j} W(r - r_j, 2h), \tag{8.6}$$

where $\mathbf{f}_j^{pressure}$ is the pressure force of the respective particle $j \in J$, $p_j$ is the particle pressure, $m_j$ is the particle mass and $\rho_j$ the particle density.

However, the pressure force is not symmetrical, hence linear and angular momentum are not conserved, violating Newton's third law [59].

*Example 8.2* Consider the interaction of two particles. The first particle considers the pressure of the second particle, whereas the second particle considers the pressure of the first particle to consider the respective pressure force $\mathbf{f}_j^{pressure}$. Since the particle pressures $p_j$ may not be equal, the pressure force is typically not symmetrical.

There are several approaches of symmetrizing the pressure force which is discussed in [59]. Here, a straightforward approach is to compute the pressure force based on the arithmetic mean of the particle pressures $p_j$.

The viscosity force $\mathbf{f}_j^{viscosity}$ depends on the strength of how viscous the fluid is, which is defined by the viscosity coefficient $\mu$. According to the SPH formulation by employing Eq. (8.1), the viscosity force thus yields

$$\mathbf{f}_j^{viscosity} = \mu \sum_{j \in J} \mathbf{u}_j \frac{m_j}{\rho_j} \nabla^2 W(r - r_j, 2h), \tag{8.7}$$

where $\mathbf{f}_j^{viscosity}$ is the viscosity force of the respective particle $j \in J$, $\mu$ is the viscosity coefficient, $\mathbf{u}_j$ is the velocity, $m_j$ is the particle mass and $\rho_j$ the particle density.

Similar to the pressure force, the viscosity force is also not symmetrical, since the velocity typically varies between particles [59].

*Example 8.3* Consider again the interaction of two particles. The first particle considers the velocity of the second particle and the second particle considers the velocity of the first particle. Since, the viscosity force depends on the velocity term, which varies between particles, the viscosity force is not symmetrical.

There exist several approaches to symmetrize the viscosity forces as discussed in [59].

### 8.1.3  External Forces

External forces include forces acting from outside of the fluid, such as gravity or buoyancy. Some of the external forces can be applied directly on the particles $j \in J$, whereas for some forces neighboring particles need to be considered by applied SPH formulations.

To that end, the gravity force is acting homogeneously on all particles $j \in J$ and is defined by

$$\mathbf{f}_j^{gravity} = \rho_j \mathbf{g}, \tag{8.8}$$

where $\mathbf{f}_j^{gravity}$ is the gravity force of the respective particle $j \in J$, $\rho_j$ the particle density and $\mathbf{g}$ the gravitational acceleration.

Buoyancy force is modeled by introducing an artificial buoyancy diffusion coefficient $b$. An artificial buoyancy force term can be defined as

$$\mathbf{f}_j^{buoyancy} = b(\rho_j - \rho_0)\mathbf{g}, \tag{8.9}$$

where $\mathbf{f}_j^{buoyancy}$ is the buoyancy force of the respective particle $j \in J$, $b > 0$ is the artificial buoyancy diffusion coefficient, $\rho_j$ the particle density and $\rho_0$ a reference density.

Such buoyancy force is essential when gaseous fluids in liquid should be modeled.

*Example 8.4* Consider an air bubble inside a liquid puddle. Here, the buoyancy force is important as it accelerates the air bubble in the direction of liquid surface.

As the buoyancy force depends on the corresponding mass-densities as defined in Eq. (8.9), large bubbles will rise faster than smaller ones.

Thus, for the simulation of complex scenarios such external forces are essential in order to provide realistic simulation results.

## 8.2 SPH Variants

There exist several variants of the SPH technique, each of which having different advantages and disadvantages as well as field of applicability. This section reviews some of the most common SPH variants. Afterwards, a more detailed treatment of the so-called *Predictive-Corrective Incompressible SPH* (PCISPH) technique is provided. PCISPH has received significant interest as it employs a dedicated predictive-corrective scheme which eventually allows to outperform other common SPH variants by up to one order of magnitude by maintaining similar accuracy.

### 8.2.1 Basic Variants

Various SPH variants have been proposed for dealing with different types of applications. These SPH variants usually differ in on how respective boundary conditions are imposed. Generally, in SPH, one of the most challenging aspects is to fulfill the incompressibility condition of fluids. Hence, different SPH variants have been introduced to enforce incompressibility employing different approaches.

In that context, two variants have been most commonly utilized: (1) *Incompressible SPH* (ISPH) which enforces incompressibility in a strong form leading to implicit formulations, and, (2) *Weakly Compressible SPH* (WCSPH) which enforces the incompressibility in a weak form leading to explicit formulations. While ISPH is generally reported to be more reliable and stable, WCSPH is typically easier to implement and applicable for parallel computations.

Beside the incompressibility term, another differentiation between different SPH variants is the applicability to fluid or solids. While both of these aforementioned SPH variants are used to model fluids, different variants have been proposed to model scenarios involving solids (e.g. bending of metal plates). In that context, the *Total Lagrangian SPH* (TLSPH) method has received significant interest, as it avoids tensile instabilities arising in conventional SPH by using a Lagrangian kernel. By that, it allows to model a range of scenarios such as crack initiation, or large deformation of solids.

In the following, these most common and widely used SPH variants, are described in more detail:

- *Weakly Compressible SPH* (WCSPH) is one of the most famous SPH variants and was initially proposed by Monaghan [57]. The main advantages of WCSPH are its ease of programming and its ability to being extended by new physical properties. Overall, WCSPH is considered one of the most well studied SPH variants.

One of the main difficulties in SPH formulations is to satisfy incompressibility. In WCSPH, a stiff equation of state is used to model pressure. Although this satisfies incompressibility, smaller time steps have to be considered in the simulation. Thus, the computational expenses increase with increasing incompressibility. As a result, WCSPH is frequently outperformed by other SPH variants, yielding one of its main disadvantages. Moreover, when applied to fluid flow applications governed by larger Reynolds numbers, it is shown that WCSPH suffers from large density variations. Thus, in such cases, smaller Mach numbers ($<0.1$) need to be employed in order to avoid nonphysical behavior in the simulation [41, 51].

For a more detailed treatment of WCPSH, the reader is referred to [4].

- *Incompressible SPH* (ISPH) is a SPH variant initially introduced by Cummins and Rudman [17] and later further improved by Shao and Lo [79]. The main idea of ISPH relies on prediction-correction fractional steps in order to enforce incompressibility. These prediction-correction fractional steps are based on a projection method that has initially been proposed in [12, 14]. However, it is shown that this projection method is suffering from an accumulation of a density error [79]. In order to overcome this problem and to further enhance the accuracy of the ISPH technique, [79] proposed a new method to enforce incompressibility. Here, an intermediate velocity is computed based on which the SPH particles are advected. Moreover, the density variation is employed as a source term as against the divergence of the intermediate velocity.

  Compared to the widely used WCSPH method, it is reported that ISPH is typically more reliable in modeling free surface flow problems [52].

- *Total Lagrangian SPH* (TLSPH) is a SPH variant proposed in [5, 99] and mainly used for modeling solids, as against the aforementioned WCSPH or ISPH method that are mainly used to model fluids. The TLSPH method mainly relies on continuum mechanics equations and constitutive models. TLSPH finds application in modeling large plastic deformations and discrete cracks. Although, standard SPH is capable of modeling large deformations, it is not suited to model cracks due to its continuum-based formulation. Moreover, TLSPH aims to overcome two main problems of the standard SPH method: (1) the tensile instability, i.e. the problem of particles frequently forming local clusters [84], and, (2) inconsistency of the SPH approximation, caused by low number of neighboring particles [40, 74]. TLSPH aims to overcome these shortcomings by employing a Lagrangian kernel and gradient correction.

Besides these variants, the *Predictive-Corrective Incompressible SPH* (PCISPH) method has received significant interest for modeling complex fluid flow scenarios. PCISPH employs a dedicated predictive-corrective scheme to enforce incompressibility to the fluid. By that, it avoids to solve the pressure Poisson equation, which eventually allows for PCISPH to outperform other SPH variants by up to one order of magnitude. Because of this major advantage, PCISPH is further considered as a baseline for optimization in Chap. 10 and, hence, explicitly reviewed in more detail next.

## 8.2.2 Predictive-Corrective Incompressible SPH

*Predictive-Corrective Incompressible SPH* (PCISPH) was initially proposed by Solenthaler and Pajarola [81] to efficiently enforce incompressibility of the fluid. Here, incompressibility is enforced by using a prediction-correction scheme, which allows to avoid the computational expenses of solving a pressure Poisson equation. By avoiding that, PCISPH allows to use a larger time step in the simulation, which typically results in an increased performance.

In PCISPH, a prediction-correction scheme is employed, where positions and velocities are temporarily forwarded in time to estimate particle densities. Based on this estimated density $\rho^*(t+1)$, the pressure $p_j(t)$ is iteratively computed for each particle $j \in J$ such that the predicted density fluctuation $\rho^*_{err}(t+1)$ is smaller than a user-defined threshold $\eta$.

More precisely, the predicted densities $\rho^*(t+1)$ are computed using the SPH density summation equation similar to Eq. (8.1), namely

$$\rho^*(t+1) = \sum_{j \in J} m_j W(r^* - r_j^*, 2h), \quad (8.10)$$

where $m_j$ is the particle mass and $r_j^*$ is the predicted particle position. $W$ is the smoothing kernel with the smoothing length $2h$ as a width. To minimize the occurring density error $\rho^*_{err}$, a correction of the current pressure is computed subsequently. Finally, the pressure force

$$F_p(t) = -m_i \sum_{j \in J} m_j \left( \frac{p_i(t)}{\rho_i^*(t)} + \frac{p_j(t)}{\rho_j^*(t)} \right) \nabla W(r^* - r_j^*, 2h) \quad (8.11)$$

is used to recompute the predicted positions and velocities for all particles $j \in J$. This procedure is repeated until it converges, i.e. $\rho^*_{err}(t+1) < \eta$. Although additional iterations are required within one time step to achieve this convergence (hereinafter referred to as *inner iterations*), PCISPH allows to use a larger time step as compared to WCSPH.

*Example 8.5* Consider the simulation of a process with a duration of $1s$, which is divided into discrete time steps. Assuming that for this process, WCSPH needs to employ time steps of e.g. $1e^{-4}s$, which yields 1000 discrete time steps. For the same process however, PCISPH allows to employ a significantly larger time step, e.g. a time step of $1e^{-3}s$, which results in 100 discrete time steps.

Therefore, PCISPH allows to significantly outperform other SPH variants such as the commonly used WCSPH. Due to these advantages, PCISPH is further considered as a baseline for novel optimization techniques as described in Chap. 10.

## 8.3  SPH and High Performance Computing

Despite all the benefits, SPH suffers from one main disadvantage which is its numerical complexity. Especially the simulation of real world phenomena frequently requires the consideration of up to hundreds of millions of particles. At the same time, the discrete particle formulation makes SPH applicable for *High Performance Computing* (HPC) methods.

As reviewed in Sect. 2.2, there are two main technologies enabling parallel execution of the SPH technique:

1. Classical CPU computations for parallelization on a single node (shared memory architecture) and multiple nodes (distributed memory architecture).
2. GPUs for massively parallel computations on a single node (single GPU architecture) and multiple nodes (multi-GPU architecture).

To that end, this section provides an overview of existing HPC solutions for SPH on both, CPUs and GPUs. Advantages and disadvantages of the corresponding techniques are discussed, as well as the open challenges.

### 8.3.1  CPU Parallelization

The parallelization on CPUs is a widely used methodology applied to a huge range of numerical algorithms. In that regard, the utilization of OpenMP for shared memory architectures as well as *Message Passing Interface* (MPI) for distributed memory architectures are considered as state of the art. There are several solutions exploiting parallelism of CPUs for particle-based methods reported in literature [24, 39, 76].

In these solutions, one of the difficulties is that the Lagrangian nature of SPH (where particles move freely in space) results in disorders of particle distributions and, hence, in scattered data in memory. However, these shortcomings are overcome by dedicated memory handling schemes and it is shown that overall these solutions lead to acceptable scalability among multiple CPUs.

This results mainly from the fact that the parallelization on a shared memory architecture (i.e. a single computer node) employing OpenMP does not require to implement a communication methodology between the computing cores. Each of the cores shares a common memory space, hence there is no need for a sophisticated communication strategy. However, such approaches have strong limitations in terms of scalability, as they only allow to utilize the computational resources of a single computer node.

Leveraging the computational resources of multiple nodes requires a communication strategy between distinct nodes in order to exchange particle information. These solutions usually employ a domain decomposition method, in which the computational domain is equally subdivided and distributed among the processors using MPI. Such approaches have the advantage of larger scalability, as they allow to utilize the computational resources of multiple computer nodes. Besides this advantage, the necessity of communication between neighboring particles renders this approach more challenging to receive optimal performance as compared to single node implementations.

However, it is shown that the parallelization of SPH on multiple nodes employing hundreds of CPUs still allows to receive close to optimal scalability. This results from the fact that the memory exchange only needs to be conducted among distinct nodes, as against individual CPUs. Hence, the communication overhead of such architectures is typically minimal.

Overall, the parallelization on CPUs is considered a standard approach that is well studied and corresponding solutions find applicability in a wide range of areas. However, these approaches on CPUs are strongly limited as compared to methods using the computational power of GPUs. While CPUs are typically composed of a few cores, GPUs allow for computations on thousands of cores. Hence, the utilization of GPUs for the parallelization of particle-based methods has gained more importance due to their larger computational power. This is described in more detail next.

### 8.3.2  GPU Parallelization

As reviewed in Sect. 2.2, GPUs allow for computations in a massively parallel manner, i.e. the execution of thousands of operations simultaneously. In particular, the sheer number of independent per-particle computations renders SPH a promising method for *General Purpose Computations on Graphics Processing Units* (GPGPU) technology. Corresponding solutions utilizing the computational power of GPUs have initially been introduced by Kolb and Cuntz [47] as well as Harada et al. [34], where the *Open Graphics Library* (OpenGL) was employed. Later, SPH implementations based on the *Compute Unified Device Architecture* (CUDA) have been developed [36].

However, in order to simulate huge domains involving millions of particles, a single GPU device is usually not sufficient anymore. In these cases, the underlying SPH implementation needs to be distributed over several devices—yielding a *multi-GPU architecture* as originally presented by Dominguez et al. [20]. Here, CUDA and MPI have been employed to parallelize the SPH simulation with up to 128 GPUs where each GPU covered the simulation of up to eight million particles. Besides that, a similar architecture has also been utilized in the solution proposed in [75].

However, how to receive optimal performance especially for huge real world scenarios remains an open question. Several aspects of the SPH method make it a highly non-trivial task to optimally utilize the computational resources of the GPUs. In fact, existing solutions suffer from severe overheads introduced by such multi-GPU architectures. Moreover, how to employ an efficient distribution of the workload of SPH simulations on multi-GPU architectures is considered one of the main challenges. In the following chapter, a detailed description on the GPU solutions proposed in this work aiming to overcome these shortcomings is provided.

# Chapter 9
# SPH on Multi-GPU Architectures

The discrete particle formulation of SPH allows for massively parallel computations on GPUs—leading to severe performance improvements as compared to classical computations on the CPU. The corresponding SPH computations can further be distributed among multiple GPUs—allowing for large scale simulations involving millions of particles.

However, as discussed in the previous chapter, obtaining optimal performance on dedicated multi-GPU architectures is a highly non-trivial task. In fact, the approaches proposed thus far suffer from severe overheads introduced by such multi-GPU architectures.

In this chapter, a dedicated multi-GPU architecture for large-scale industrial SPH simulations is presented (originally proposed in [92]). To that end, an advanced load balancing scheme for SPH simulations is proposed which addresses the inherent overheads introduced by these multi-GPU architectures. The proposed scheme frequently conducts a so-called *domain decomposition correction*, which adjusts the current distribution of the workloads among the GPUs. Since this frequently requires re-allocations of memory, dedicated memory handling schemes are introduced. Experimental evaluations with an industrial SPH solver confirm that sophisticated memory handling allows to reduce the synchronization overhead and, therefore, results in an increased performance.

In the remainder of this chapter, the solution is described in detail as follows: Sect. 9.1 describes the background of multi-GPU architectures used for the purpose of large-scale simulations. Afterwards, Sect. 9.2 provides details on the proposed advanced load balancing scheme. Finally, Sect. 9.3 summarizes the obtained results from the experimental evaluations before this chapter is concluded in Sect. 9.4.

© The Author(s), under exclusive license to Springer Nature Switzerland AG 2021
K. Verma, R. Wille, *High Performance Simulation for Industrial Paint Shop Applications*, https://doi.org/10.1007/978-3-030-71625-7_9

## 9.1   Background

This section discusses the background of SPH simulations on corresponding multi-GPU architectures. To this end, a comprehensive description of the underlying methodologies of multi-GPU architectures is provided. Afterwards, the motivation for a sophisticated load balancing scheme is discussed.

### 9.1.1   Basic Architecture

The performance of any SPH simulation strongly depends on the fact that neighboring particles need to be frequently accessed during one computational iteration. More precisely, as reviewed in Sect. 8.1 and defined by Eq. (8.1), the scalar quantity $A$ is interpolated by a weighted function of all particles which are located within the influence radius $2h$ (defined by the smoothing length). Applying a straightforward nearest neighbor search algorithm, this requires the iteration through the entire fluid domain—yielding a complexity of $O(|J|^2)$. Although polynomial, the sheer number $|J|$ of particles makes this straight-forward approach infeasible for many practically relevant problems (e.g. in applications such as simulations of wave interactions with an off-shore oil rig platform, more than one billion particles have to be considered [78]).

Hence, corresponding optimizations have been introduced which rely on a so-called *virtual search grid* as illustrated in Fig. 9.1. Here, the entire fluid domain is divided into a search grid where each cell has the size of the influence radius $2h$. By this, it can be guaranteed that, for a considered particle $j \in J$ (exemplarily denoted by a red dot in Fig. 9.1), all neighboring particles must be located within the adjacent cells. This way, instead of iterating through the entire fluid domain, only a subset of it (highlighted grey in Fig. 9.1) has to be considered in order to determine the neighboring particles (denoted by orange dots in Fig. 9.1).

This search grid not only allows for a fast nearest neighbor search, but also provides a scheme how to divide the entire fluid domain into sub-domains. For

**Fig. 9.1** Virtual search grid

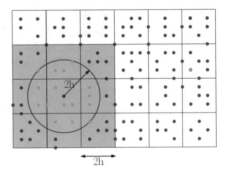

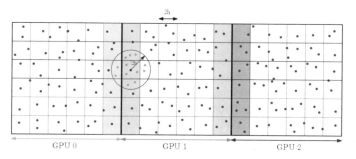

**Fig. 9.2** Subdomain distribution for three GPUs

example, a spatial subdivision based on the grid cells could be applied—leaving every sub-domain with an equal number of cells of the grid. These sub-domains can then be distributed to the corresponding devices on a multi-GPU architecture.

The basic SPH multi-GPU solution developed in this work employs such a spatial subdivision of the domain to partition the whole domain into individual subdomains. These subdomains are distributed to the corresponding GPUs and executed in parallel. Figure 9.2 illustrates a subdivision into three subdomains. Generally, this method yields a setup where each subdomain has two neighboring subdomains, except for those at the perimeter of the domain, which have only one neighbor. However, the boundaries of each subdomain need special consideration on a multi-GPU architecture as illustrated by the following example.

*Example 9.1* Consider a particle $j \in J$ at the perimeter of one centered sub-domain (see Fig. 9.2). The neighbors of $j$ within $2h$ are not only located in cells of its own subdomain (covered by GPU 1 in Fig. 9.2), but also in cells of the neighboring subdomain (covered by GPU 0). Since the subdomains are distributed to different GPUs, the neighbors of $j$ are located in a distinct device and, hence, a different memory pool. This hinders fast neighbor access.

In order to accelerate neighbor access, each GPU should therefore hold a copy of the data located at the *edge* of its adjacent subdomains, i.e. all cells within $2h$ at the perimeter of a subdomain. These edges are also referred to as *halo* of a subdomain.

More precisely, after every particle modification, a brief synchronization over all halos takes place. Afterwards, the parallel computation on all subdomains can continue with the updated values.

## 9.1.2 Motivation

In general, the goal for every application executed on a multi-GPU architecture is to split the workload in a way which allows each GPU to complete the respectively assigned computations in the same amount of time. Since such a behavior is usually difficult to guarantee, the alternate goal is to minimize the gap between the longest and shortest time consumption.

In the case of SPH simulations, the key to achieve optimal performance is to determine the optimal positions for a subdivision of the domain. However, for that purpose, important characteristics of SPH need to be considered.

First, although the partitioning of the entire fluid domain into sub-domains (defined by the grid cells) provides a scheme how to distribute the corresponding workload over the respective GPU devices, it is not always useful to apply a spatial subdivision based on the geometry of the domain only. In fact, particles usually do not equally distribute along the domain—making a subdivision based on the geometry unbalanced.

*Example 9.2* Consider a scenario, where a dam break has to be simulated using SPH techniques. At time step $t = 0$, all particles are clustered at one side of the domain, namely the water reservoir. At the end of the simulation (after the dam broke), most particles are likely been distributed equally among the domain.

Therefore, a subdivision according to the distribution of particles is essential for most applications.

Besides that, the inherent moving nature of the particles means that particles do not stay within "their" respective grid, but freely move in space—requiring to frequently update the corresponding workload of the respective GPUs. Overall, these characteristics make load balancing a complex task and determining a sophisticated load balancing strategy is the key to achieve optimal performance.

Unfortunately, rather few works exist on this subject thus far. One solution has been presented in the work of [75]. Here, a simple a posteriori load balancing system is introduced, which shows high robustness in the performed test cases. However, the proposed load balancing introduces an overhead, mainly caused by moving edges of one GPU to another. In the work of [20] a similar method is presented—yielding the same disadvantages. Besides that, load balancing for particle-based simulations on multi-GPU architectures has also been considered for classical *Molecular Dynamics* (MD, see e.g. [89, 100]). But MD differs from SPH simulations and usually deals with smaller motions and a wider variety of particles—making the load balancing tasks significantly different. Hence, how to employ an efficient distribution of the workload of SPH simulations to the respective GPUs remains an open question.

## 9.2 Advanced Load Balancing

In this section, an approach for an advanced distribution of the SPH simulation workload to GPU architectures is proposed. To this end, first the general idea of the proposed solution is outlined. Afterwards, important implementation details are covered, which need to be considered in order to minimize the overhead.

### 9.2.1 General Idea

In any SPH implementation, many unpredictable factors, such as the fluid movement, may influence the computation time during run-time. Hence, an advanced a
posteriori load balancing methodology needs to be chosen. In our solution, first an
initial decomposition of the domain is calculated. More precisely, the initial number
of particles per GPU is approximated by

$$N_i = \frac{N_t}{N_g}, \tag{9.1}$$

where $N_i$ is the initial number of particles per GPU, $N_t$ the total number of particles,
and $N_g$ the number of GPUs.

This distribution is used to compute the first time steps, while recording the
amount of time spent by each GPU. After an update interval of $n$ time steps, the
algorithm iterates through the time data collected from each GPU. Then, the *relative
runtime difference* $d_r$ between two neighboring devices $g_i$ and $g_{i+1}$ is determined
by

$$d_r = \frac{r_{i+1} - r_i}{r_i}, \tag{9.2}$$

where $d_r$ is the relative runtime difference, $r_i$ the runtime of $g_i$, and $r_{i+1}$ the runtime
of $g_{i+1}$.

If $d_r$ is larger than a threshold factor $p$, a *domain decomposition correction* from
$g_i$ to $g_i + 1$ is applied, i.e. one *edge* (as introduced in Sect. 9.1) is moved from the
GPU with the longer execution time to the GPU with the shorter one. If $d_r$ is smaller
than $-p$, a *domain decomposition correction* from $g_i + 1$ to $g_i$ is applied. This
process of corrections is repeated every $n$ further time steps again. Algorithm 11
explains this procedure in pseudocode.

*Example 9.3* Consider the simplest case for a *domain decomposition correction*:
a multi-GPU architecture composed of two GPUs as illustrated in Fig. 9.3. GPU
$g_i$ required only eight seconds to compute $n$ time steps, while GPU $g_{i+1}$ required
ten seconds. Hence, a relative time difference of $d_r = 0.25$ results—yielding an
unbalanced state. In order to balance that state, one *edge* of the virtual search grid is
shifted from $g_{i+1}$ to $g_i$.

The basic concept of this load balancing strategy is rather straight-forward.
However, taking a more detailed look into GPU and CUDA architectures, applying
such a domain decomposition correction with a minimum amount of overhead is a
non-trivial task. In fact, in our implementation each GPU internally stores its particle
data consecutively in an array. Within that array, the particle data are kept sorted, in
order maintain cell positions of the search grid. Hence, particles at the perimeter of
the domain are also kept at the perimeter of the array. This requires re-allocations
of the respective array memories when a domain decomposition correction (and,

**Algorithm 11** Balance

$n \leftarrow$ `update interval`
$p \leftarrow$ `threshold`
$K \leftarrow$ `time steps`
$N_g \leftarrow$ `number of GPUs`
**for** $k \in \{1, \ldots, K\}$ **do**
    `Calculate Time Step`
    **if** $k \mod n = 0$ **then**
        **for each** $GPU g_i \in \{g_0, \ldots, g_{N_g}\}$ **do**
            **if** $RuntimeDifference(g_i, g_{i+1}) > p$ **then**
                `Decomposition Correction`$(g_i, g_{i+1})$
            **end if**
            **if** $RuntimeDifference(g_i, g_{i+1}) < -p$ **then**
                `Decomposition Correction`$(g_{i+1}, g_i)$
            **end if**
        **end for**
    **end if**
    `Sort Particles`
**end for**

**Fig. 9.3** Unbalanced state in a system composed of two GPUs

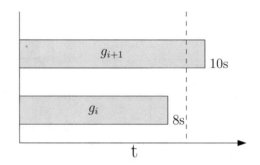

hence, a re-balancing) and an *exchange of halos* is conducted—accomplished by a *cudaMalloc*-call in the considered CUDA architecture. Unfortunately, this is a synchronous call which causes all GPUs to synchronize. Overall, this results in a major loss in performance.

*Example 9.4* Consider again a multi-GPU architecture composed of two GPUs and assume that data is currently allocated in arrays as illustrated in Fig. 9.4a. Now, the particles of the left edge of the sub-domain covered by $g_{i+1}$ (stored in the front of the array of $g_{i+1}$) should be shifted to the right edge of the sub-domain covered by $g_i$. This requires the allocation of new memory to the array storing the particles of $g_i$.

This, however, significantly affects the execution of the respective computations as illustrated in Fig. 9.4b. Here, kernel functions (denoted by $K_{i_0}$, $K_{i_1}$, ...) are executed in parallel among both GPUs, until cudaMalloc() causes synchronization. Then, all computations have to pause in order to synchronize. Assuming larger multi-GPU systems composed of hundreds of GPUs which frequently need to

**Fig. 9.4** Overhead caused by re-allocations. (**a**) Re-allocation of memory during a decomposition correction. (**b**) Resulting synchronization step

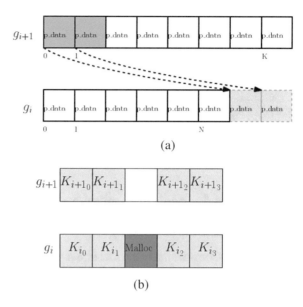

allocate new memory, this easily accumulates to a severe number of synchronization steps in which all GPUs (even if they are not affected) have to pause—a significant loss in performance.

Obviously, such a loss in performance should be avoided. For that purpose, two different approaches are proposed which are introduced and discussed in the next sections.

### 9.2.2   Using Internal Cache

A first solution employs an additional local cache to shift data (as illustrated in Fig. 9.5). Here, *cudaMemcpyAsync* from the CUDA API is used, which allows for an efficient asynchronous copy of data from the local data array to a local cache. After copying, the pointers between the array and the cache are swapped—leaving the local array with the reduced data and the cache with the original objects. During that process, the capacity of both, the data array and cache is only reduced when the capacity is significantly larger than the used memory. Instead, only the internal number of elements is modified without actually freeing any memory. The benefit of that method is the minimized synchronization caused by allocating new memory, since *cudaMalloc* only needs to be called when the amount of data to be processed in one GPU is larger than the initial amount $N_i$ of particles covered per GPU. This drastic reduction of synchronization efforts results in a noticeably improved performance.

**Fig. 9.5** Avoiding
synchronization using
internal cache

**Fig. 9.6** Avoiding
synchronization using
pointers

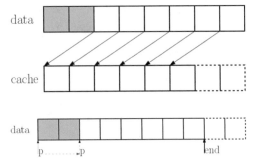

In case new data needs to be shifted back to the device, the local data array is examined if the new data fits to the already allocated data chunk. If that check fails, new memory needs to be allocated.

*Example 9.5* Consider again the scenario discussed in Example 9.4. Employing the cache implementation as described above, the capacity of $g_{i+1}$ array is not reduced. Instead, data is just shifted by one block. Assuming that one edge of $g_i$ needs to be shifted back in the next domain decomposition correction, e.g. due to unpredictable fluid structure behavior, memory allocation can be avoided since data has not been freed during the previous domain decomposition correction.

The obvious drawback of this method is the demand for copying pointers and the increase of memory consumption. However, as previously discussed, particle data always needs to be kept sorted. For that purpose, a radix sort is employed, which is one of the fastest sorting algorithms on GPUs. Since radix sort is typically an out-of-place sorting algorithm, a temporary storage buffer is required. Therefore, the memory consumption of radix sort is $O(2N + P)$, where $N$ is the length of the input data to be sorted and $P$ the number of streaming multiprocessors on the device. For that purpose, the cache can be employed as a temporary storage, which leaves the high watermark memory consumption unchanged by radix sort. However, in case memory requirement is the bottleneck for a given computation, a second solution with reduced memory consumption is proposed next.

### 9.2.3  Using Pointers

The main drawback of the cache method is the increased memory consumption. To avoid that memory overhead, an alternative method is proposed in which data is not shifted using an additional cache, but by shifting pointers to the beginning of the data block (as illustrated in Fig. 9.6). For example, in case the first two elements should be deleted, the pointer is incremented accordingly.

Of course, data is not only removed from the front but also from the back, when data needs to be transferred from $GPU_i$ to $GPU_{i+1}$. In that case, the pointer to end is moved accordingly.

Similar to the cache implementation, no memory needs to be freed—leaving the capacity of the array unchanged and avoiding frequent memory allocations. However, in contrast to the cache implementation, free memory blocks can occur in the front or in the back of the array when the pointer-solution is applied. Because of this, there are frequent checks for unoccupied space in the front and in the back of the array which can be used to accordingly add data. Only when there is not enough free space available, i.e. the amount of needed memory is larger than the original setup $N_i$, new memory is allocated.

Clearly, the advantage of this method is that internal cache can be omitted, which results in a decreased memory consumption. However, the inherent out-of-place behavior of radix sort occupies a temporary storage buffer for sorting. Therefore, the high watermark memory consumption during sorting is defined as $O(2N + P)$. Efficient in-place sorting methodologies on GPUs do not provide comparable time complexity. For example, comparing the so-called *bitonic sort* for GPUs as initially proposed by Peters et al. [72]. Bitonic sort is an efficient comparison-based in-place sorting algorithm for CUDA. It offers a time complexity of $O(N \log^2 N)$ compared to a complexity of $O(N \log N)$ for radix sort. Hence, the decreased memory consumption is traded off by a weaker sorting performance.

## 9.3   Experimental Evaluations

In order to evaluate the performance of the proposed load balancing approaches for the SPH technique, the methods described above have been implemented in C++ for an industrial SPH solver. Experiments with both, an academic dam break scenario and an industrial spray wash scenario have been conducted, whose results are summarized in this section. In the following, the experimental setup is described in detail.

### 9.3.1   Experimental Setup

In the conducted experiments, the cache and the pointer implementation (as introduced in Sects. 9.2.2 and 9.2.3, respectively) with respect to speedup and efficiency are evaluated.

The speedup is defined as

$$S = \frac{T_s(n)}{T_p(n, p)} \tag{9.3}$$

where $S$ is the speedup, $n$ the size of the input (number of particles in the system), $T_S$ the execution time of the single GPU implementation, $T_p$ the execution time of the multi-GPU implementation, and $p$ the number of used GPUs. The efficiency is defined as

$$E = \frac{S}{p} \qquad (9.4)$$

where $E$ is the efficiency, $S$ the speedup, and $p$ the number of GPUs.

All evaluations have been conducted on GPU systems composed of Nvidia GTX 1080 Ti, which contain 3584 CUDA cores with a memory bandwidth of 484 GB/s. The source code was compiled on Ubuntu v16.04 using gcc v4.5.3 and the CUDA Toolkit v9.1. The experiments include an industrial spray wash scenario as well as an academic dam break test case, as described next.

### 9.3.2  Dam Break Simulation

The first test case considers a dam break scenario with a rotating obstacle in the middle as sketched in Fig. 9.7. This scenario is especially suited to evaluate the proposed load balancing methodology, as here, the particle distribution changes significantly as the simulation progresses. Hence, a sophisticated load balancing scheme is key to obtain good performance. To evaluate the scalability of the method, this scenario has been considered using 385k, 2.38mio, and 5.26mio particles.

Table 9.1a shows the speedup and efficiency using the cache implementation. The values show that with larger number of particles, the achieved speedup increases. This is inherently the case, since the more particles that need to be computed the smaller the proportion spent on communication between GPUs.

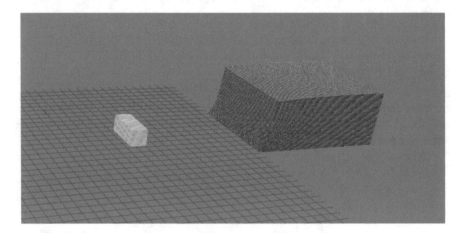

**Fig. 9.7**  Sketch of the considered scenario (dam break)

**Table 9.1**  Results obtained by the experimental evaluation

| Speedup | | | | Efficiency | | |
|---|---|---|---|---|---|---|
| # GPUs | 130k | 2,40mio | 5.26mio | 130k | 2,40mio | 5.26mio |
| (a) Using the cache implementation from Sect. 9.2.2 | | | | | | |
| 1 GPU | 1 | 1 | 1 | 1 | 1 | 1 |
| 2 GPU | 1.56 | 1.64 | 1.68 | 0.78 | 0.82 | 0.84 |
| 3 GPU | 2.05 | 2.25 | 2.31 | 0.68 | 0.75 | 0.77 |
| (b) Using the pointer implementation from Sect. 9.2.3 | | | | | | |
| 1 GPU | 1 | 1 | 1 | 1 | 1 | 1 |
| 2 GPU | 1.61 | 1.70 | 1.79 | 0.81 | 0.85 | 0.90 |
| 3 GPU | 2.14 | 2.36 | 2.45 | 0.71 | 0.79 | 0.82 |

Table 9.1b shows the speedup and efficiency using the pointers implementation. The values show that in the present test example, the pointer implementation is superior in all setups compared to the cache implementation. This is mainly because the dam break is an asymmetrical problem, where particles are clustered at one side in the beginning of the process and equally distributed in the end. Therefore, frequent communication between GPUs is necessary, since particles move quickly between the respective sub-domains. This is suitable for the pointer implementation which needs less overhead by copying pointers to local cache and, hence, results in an improved speedup.

The presented results also show that the methods proposed in this work, scale well for smaller number of particles. For example, an efficiency of 0.79 is achieved when using 2,4mio particles on three GPUs, but also for a tiny number of particles of 130k on three GPUs, an efficiency of 0.71 is achieved.

## 9.3.3  Spray Wash Simulation

As an industrial test case, a spray wash scenario of the automotive industry is considered. This process is part of the pre-treatment stage of the automotive paint shop. Here, assemblies or entire car bodies are cleaned by external liquid sprayed onto the objects by nozzles. These nozzles are arranged in a defined pattern, to spray water (or cleansing fluid mixtures) onto the surfaces. By that, dirt on the surface or similar defects should be eliminated, to subsequently ensure a clean paint process.

The case considered in this work, includes an automotive door part as sketched in Fig. 9.8a. The nozzles all have the same geometry and flow pattern to ensure a uniform liquid distribution. Throughout the process, the door will move along the $x$-axis from one end to the other with a length of 2.5 m. The line speed of the door is defined as $u = 0.2$ m/s. The nozzles induce water with a density of $\rho = 1000\,\text{kg/m}^3$ and a kinematic viscosity of $v = 8.94 \times 10^{-7}\,\text{m}^2/\text{s}$. Each of the nozzles induces the water with a flow rate of $V = 20\,\text{L/min}$.

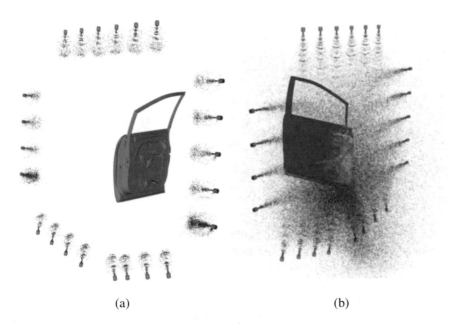

(a)                                                          (b)

**Fig. 9.8** Spray wash simulation scenario with an automotive door part. (**a**) Setup of the spray wash scenario. (**b**) Simulation of the spray wash scenario at $t = 2$ s. The color coding indicates the contact time

**Table 9.2** Speedup and efficiency obtained for the spray wash scenario

| # GPUs | Cache implementation | | Pointer implementation | |
|---|---|---|---|---|
|  | Speedup | Efficiency | Speedup | Efficiency |
| 1 GPU | 1 | 1 | 1 | 1 |
| 2 GPU | 1.71 | 0.86 | 1.75 | 0.88 |
| 3 GPU | 2.32 | 0.77 | 2.49 | 0.83 |
| 4 GPU | 2.95 | 0.74 | 3.12 | 0.78 |

Initially, the total particle count of the simulation is 3mio which are all solid particles defining the geometry of the door part. As the simulation progresses, the nozzles induce water particles, hence, the total number of particles increases with time. The bottom of the domain is defined as a *sink* boundary condition, i.e. particles in the bottom of the domain will be deleted from the simulation.

Figure 9.8b shows the simulation at time step $t = 2$ s. The color coding on the surfaces shows the contact time i.e. the rinsing duration of the liquid. The performance of the simulation is again evaluated in terms of speedup and efficiency.

Table 9.2 shows the obtained speedup and efficiency obtained for both, the cache and the pointer implementation.

It is shown that overall the pointer implementation yields superior efficiency as compared to the cache implementation. In this scenario, this is mainly caused by the moving door part, which requires frequent redistribution of particles from one subdomain to the other. Moreover, the sinks on the bottom of the domain, frequently delete particles which again augments the asynchronous nature of this scenario.

Overall, the presented results show that the proposed load balancing schemes are also applicable to industrial problems. For example, an efficiency of 0.88 is achieved for two GPUs using the pointer implementations. Using four GPUs, the efficiency is still 0.78, rendering this implementation effective for an industrial use case with a relatively small number of particles.

## 9.4  Summary

In this chapter, a load balanced multi-GPU CUDA-based implementation of an industrial SPH solver is presented. To this end, a sophisticated domain decomposition strategy is employed and a load balancing methodology is proposed, which also scales well for smaller number of particles. Important implementation details based on the CUDA architecture are provided, which reduce the synchronization overhead and, therefore, result in an increased efficiency. Future work includes the implementation of a multi-node solution, hence, the introduction of an additional level of parallelization, and further optimization on the domain decomposition strategy to achieve further speedups.

# Chapter 10
# SPH Variants on Multi-GPU Architectures

There exist several variants of the SPH method, each of which yielding different advantages and disadvantages. One promising variant is the *Predictive-Corrective Incompressible SPH* (PCISPH) method as reviewed in Sect. 8.2.2. It allows to use a much smaller time step, hence it typically outperforms other SPH variants by up to one order of magnitude.

However, one of the main disadvantages of this variant is that it does not provide an obvious parallelization scheme as e.g. the well studied *Weakly Compressible SPH* (WCSPH, as reviewed in Sect. 8.2) method. This is essentially caused by the increased synchronization effort, which thus far prevents PCISPH from exploiting state of the art *High Performance Computing* (HPC) architectures. This is a serious drawback, as efficient parallelization is considered essential to further improve the scalability of such approaches and thereby make them suitable for the simulation of real-world phenomena in appropriate resolution—frequently taking into account hundreds of millions of particles.

In this chapter, a PCISPH implementation is presented that is suitable for distributed multi-GPU architectures (originally proposed in [93, 97]. In order to accomplish that, a PCISPH workflow is presented that allows for large hardware utilization. To that end, dedicated load balancing and optimization techniques are presented, which eventually allow to fully exploit the advantages of the PCISPH method. The proposed methods are implemented on top of an industrial PCISPH solver, and, for the first time, applied to model an industrial engineering scenario. Experimental evaluations confirm that, by using the proposed methods, PCISPH can be efficiently applied to model real world phenomena.

In the remainder of this chapter, the solution is described in detail as follows: Sect. 10.1 describes the state of the art and limitations of PCISPH. Afterwards, Sect. 10.2 provides details on the proposed distributed multi-GPU scheme. Section 10.3 provides detailed descriptions on the proposed optimization techniques and, finally, Sect. 10.4 summarizes the results from experimental evaluations, before the conclusion is drawn in Sect. 10.5.

K. Verma, R. Wille, *High Performance Simulation for Industrial Paint Shop Applications*, https://doi.org/10.1007/978-3-030-71625-7_10

## 10.1  Background

One of the main difficulties of SPH is the fulfillment of the incompressibility conditions of liquids. For that purpose, *Weakly Compressible SPH* (WCSPH) was proposed by [57], which has become one of the most popular variants of SPH. As reviewed in Sect. 8.2, in WCSPH a stiff equation of state is used to model the pressure. Although this satisfies the incompressibility, WCSPH still suffers from a strong time constraint. In order to enforce a higher incompressibility, smaller time steps have to be considered in the simulation. Thus, the computational expenses increase with increasing incompressibility. As a result, WCSPH is frequently not feasibly to model complex scientific or industrial fluid flow scenarios.

To address this shortcoming, *Predictive-Corrective Incompressible SPH* (PCISPH) was proposed in [81]. In PCISPH, incompressibility is enforced by employing a prediction-correction scheme, which allows to avoid the computational expenses of solving a pressure Poisson equation, as reviewed in Sect. 8.2.2. Through this scheme, PCISPH allows the use of a much larger time step as compared to WCSPH. This results in a significantly improved performance, significantly outperforming the commonly used WCSPH ([71, 81]).

Similar to the WCSPH method, the discrete particle formulation of physical quantities also makes the PCISPH method generally suitable for parallelization—especially for *General Purpose Computations on Graphics Processing Units* (GPGPU) technology. However, certain characteristics of PCISPH result in an exceeding synchronization effort especially on distributed multi-GPU architectures.

As described in Sect. 9.1, multi-GPU solutions employ a spatial subdivision of the domain to partition the whole domain into individual subdomains. These subdomains are distributed to the corresponding GPUs and executed in parallel. After every particle modification, a brief synchronization over all edges (also referred to as *halo*) of a subdomain is conducted. Afterwards, the parallel computation on all subdomains can continue with the updated values. Overall, this allows for significant speed-ups e.g. for WCSPH due to parallelization on multiple GPUs.

This synchronization over all halos constitutes one major bottleneck of any SPH parallelization method using a domain decomposition technique. But while e.g. WCSPH still gains significant total improvements from that, it constitutes a "showstopper" for PCISPH. This is caused by the predictive-corrective scheme employed in PCISPH, which requires a highly dedicated synchronization scheme.

In order to discuss this explicitly, consider a PCISPH implementation based on the formulations described in Sect. 8.2.2. To enforce incompressibility, a prediction-correction scheme to determine the particle pressures is employed. This scheme requires an additional iteration loop within every time step computation, in which the new particle densities are estimated. This loop is iterated until the convergence is achieved. During this process, the discrete particle values are frequently modified, which requires a significant amount of halo exchange processes. In total, this results in $2 + K \cdot 4$ halo exchanges, where $K$ refers to the number of iterations until convergence is achieved (typically 3–5 iterations).

This synchronization effort specifically yields a bottleneck when distributed memory architectures are used, since here, the limited memory bandwidth increasingly restrains the performance of these memory exchanges. This inherent synchronization requirement, thus far strongly limits the applicability of PCISPH, since parallel architectures are an essential technique to further improve the scalability of particle-based methods and, by that, make them suitable to model real world phenomena.

## 10.2 Distributed Multi-GPU Architecture

In this section, the proposed multi-GPU PCISPH architecture for heterogeneous GPU clusters is discussed. To this end, at first the basic methodologies of the proposed solution is outlined.

Algorithm 12 shows the general multi-GPU implementation of PCISPH. Here, as discussed above, inner iterations are used to enforce the incompressibility by a prediction-correction scheme to determine the particle pressures. The velocities and positions are temporarily forwarded in time and used to estimate the new particle densities (see line 14). For each particle $j \in J$, the predicted variation from the reference density is computed and used to compute the pressure values, which are then used for the computation of the pressure force (see lines 15–18). This process is iterated until it converges, i.e. until all particle density fluctuations are smaller than a defined threshold $\eta$. Since in every iteration the particle values are frequently updated, the halos need to be exchanged between GPUs after every particle modification.

Hence, in distributed multi-GPU PCISPH simulations, the halos of each subdomain need to be frequently exchanged between GPUs in order to ensure correct particle values. These exchanges of halos inherently constitute one of the major bottlenecks in multi-GPU architectures, due to the massive communication burden. Compared to other variants, the predictive-corrective scheme of PCISPH results in a larger number of these synchronization points, rendering this variant significantly more challenging to obtain optimal speedups. Hence, one focus of the approach used in this work is on optimizing these data exchanges.

In the proposed method the exchange of halos has been implemented using *CUDA-aware MPI* (Message Passing Interface, [27, 31]) with support of *Unified Virtual Addressing* (UVA, [77]). This allows a single address space for all CPU and GPU memory. Therefore, the particle memory can be transferred directly between GPUs (peer-to-peer) without the need to stage the data through CPU memory. By employing this technology, the communication burden is completely offloaded from the CPU to the GPU, which allows to develop highly asynchronous code.

Moreover, the proposed PCISPH implementation uses fully device-resident data structures and all the computations are done merely on GPUs, exploiting the full

---

**Algorithm 12** Parallel PCISPH

---

1:  $T \leftarrow$ time steps
2:  $N_g \leftarrow$ number of GPUs
3:  $\eta \leftarrow$ maximum allowed density error
4:  **for** $t \in \{0, \ldots, T\}$ **do**
5:      **for** $GPU g_i \in \{g_0, \ldots, g_{N_g}\}$ **do**
6:          Update Boundary Particles
7:          **ExchangeHalo**$(g_i, g_{i+1})$
8:          Compute $F^{v,g,ext}$
9:          **ExchangeHalo**$(g_i, g_{i+1})$
10:         **while** $\rho^*_{err} > \eta$ **do**
11:             Update Boundary Particles
12:             **ExchangeHalo**$(g_i, g_{i+1})$
13:             Predict Density $\rho^*(t+1)$
14:             Predict Density Variation $\rho^*_{err}(t+1)$
15:             Compute Pressure $p(i) + = f(\rho^*_{err}(t+1))$
16:             **ExchangeHalo**$(g_i, g_{i+1})$
17:             Pressure Correction
18:             **ExchangeHalo**$(g_i, g_{i+1})$
19:             Compute Pressure Force $F_p(t)$
20:             **ExchangeHalo**$(g_i, g_{i+1})$
21:         **end while**
22:         Compute Velocity $v(t+1)$
23:         Compute Position $x(t+1)$
24:     **end for**
25: **end for**

---

potential of massively parallel devices. The CPUs act as drivers, which mainly prepare the employed data structures before the exchange of halos takes place. More precisely, the CPUs perform the following tasks:

1. Update and mark the halo and padding regions in the virtual search grid to preparing exchange of data.
2. Determine destination and source ranks for every halo/padding and compute corresponding data offsets.
3. After data exchange, update halos and paddings in the virtual search grid.

Every other computational burden is exclusively done on GPUs, such as all the simulation specific computations (e.g. computing particle interaction).

By exploiting this technique, an accelerated peer-to-peer communication between GPUs can be achieved. However, further optimization is required to achieve satisfying performance for multi-GPU PCISPH on heterogeneous clusters.

## 10.3  Optimization Techniques

Receiving optimal performance for PCISPH simulations on multi-GPUs on heterogeneous clusters is a non-trivial task. The inherent moving nature of particles introduces many obstacles, which are not present in classical grid-based computational fluid dynamics methods. In order to accelerate PCISPH in massively parallel architectures, several optimizations are introduced as discussed in this section.

### 10.3.1 Load Balancing

The inherent moving nature of the particles means that particles do not stay within "their" respective grid, but freely move in space—requiring to frequently update the corresponding workload of the respective GPUs.

For that purpose, the load balancing methodology as discussed in Sect. 9.2 is utilized and further modified for distributed memory architectures. In this solution, first an initial decomposition of the domain is calculated. More precisely, the initial number of particles per GPU is approximated by

$$N_i = \frac{N_t}{N_g},\tag{10.1}$$

where $N_i$ is the initial number of particles per GPU, $N_t$ the total number of particles, and $N_g$ the number of GPUs.

This distribution is used to compute the first time steps, while recording the amount of time spent by each GPU. After an update interval of $n$ time steps, the algorithm iterates through the time data collected from each GPU. Then, the *relative runtime difference* $d_r$ between two neighboring devices $g_i$ and $g_{i+1}$ is determined by

$$d_r = \frac{r_{i+1} - r_i}{r_i},\tag{10.2}$$

where $d_r$ is the relative runtime difference, $r_i$ the runtime of $g_i$, and $r_{i+1}$ the runtime of $g_{i+1}$.

If $d_r$ is larger than a threshold factor $p$, a *domain decomposition correction* from $g_i$ to $g_i + 1$ is applied, i.e. one *edge* is moved from the GPU with the longer execution time to the GPU with the shorter one. If $d_r$ is smaller than $-p$, a *domain decomposition correction* from $g_i + 1$ to $g_i$ is applied. This process of corrections is repeated every $n$ further time steps.

This domain decomposition correction is thereby completely conducted on GPUs, i.e. the particle memory is not staged through the host. Instead, memory is shifted directly from $g_i$ to its neighboring GPUs. The only workload done on CPU is to update the corresponding pointers to the beginning of the subdomains and corresponding halos and paddings. This approach yields the advantage that no memory needs to be staged through the host, which typically is a major bottleneck. Inherently, this suggests that these steps cannot be performed asynchronously to any computations done on GPU, as the whole particle layout is changed during

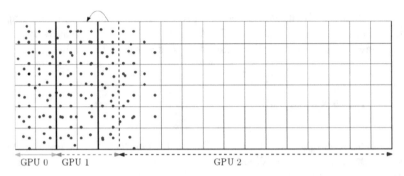

**Fig. 10.1** Load balancing step in a three GPU setup

this approach. Moreover, outsourcing these steps to the CPU would not result in a performance increase since, again, the bottleneck is to stage the memory from the host to the device. Hence, using such fully GPU resident approach is superior to any kind of approach where the workload is executed on the CPU.

Figure 10.1 shows a load balancing step for a three GPU setup. The particles are moving from left to right, hence the workload is moving towards GPU 2. In order to keep the workload balanced, one edge of GPU 0 is moved to GPU 1.

As a result of this methodology, the workload is equally distributed among the different GPUs. Due to the a posteriori load balancing, which is based on the actual execution time of the GPU kernel, this method is especially suited for heterogeneous clusters, since it takes differences in computational hardware into account.

Besides the inherent moving nature of the particles, the memory transfers introduced by the necessity to exchange halos is another important aspect to consider for optimization.

### 10.3.2   Overlapping Memory Transfers

In any SPH simulation on multi-GPU architectures, memory needs to be exchanged between GPUs after particle data has been modified—yielding the bottleneck discussed above. Here, these synchronization problems are overcome by conducting the required halo exchanges (and, by this, the communication between GPUs) in parallel to the actual computations. To this end, a revised workflow for PCISPH is utilized which is illustrated in Fig. 10.2.

Here, the computation is essentially divided into three separate tasks:

1. First, all computations within the halo regions are conducted. For the subdomains located in the perimeter of the domain, the computations for only one halo region need to be executed, i.e. the right halo for the leftmost subdomain (hereinafter referred to as $halo_r$) and the left halo for the rightmost subdomain (hereinafter

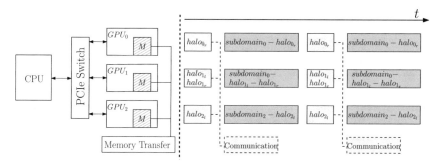

**Fig. 10.2** Overlapping memory transfers with actual computations

referred to as $halo_l$). For the subdomains located in the middle of the domain, the computations for both, $halo_l$ and $halo_r$, are conducted.

2. Second, the computations for the remainder of the subdomains are performed. For the leftmost and rightmost subdomains, this yields the computations within the region $subdomain\text{-}halo_r$ and $subdomain\text{-}halo_l$, respectively. For the subdomains located in the middle of the domain, the computations within the region $subdomain\text{-}halo_r\text{-}halo_l$ are executed.

3. Finally, the halos between the subdomains are exchanged, i.e. the memory transfers are conducted. Since the interactions within the halo regions have already been computed, this can now be conducted in parallel to the second task.

Overall, this re-arrangement of tasks reduces the synchronization problems of existing parallelization schemes for SPH. While this also could help to improve existing SPH solutions such as WCSPH, it particular overcomes the bottleneck of PCISPH solutions and, eventually, allows for a more efficient memory transfer and, hence, parallelization for this scheme.

However, it is shown that the time spent on memory transfers is frequently larger than the time spent on the actual computation. Hence, the memory transfers cannot be completely hidden behind the computations. In order to improve on this aspect, further effort includes the optimization of these memory transfers by an optimized particle data representation.

### 10.3.3  Optimizing Particle Data Representation

Further optimization introduced to the multi-GPU PCISPH method is related to the data structures representing the particle values. For that purpose, *structure of arrays* (SOA) are employed to group certain quantities in order to efficiently organize the memory of the particle data. This essentially allows to only exchange the physical properties of particles which have been modified by the preceding operation. More precisely, separate arrays are kept for individually storing:

1. Positions and pressure.
2. Velocities and density.
3. Accelerations and pressure coefficient.
4. Id, kinematic viscosity and mass.
5. Gamma, initial density and speed of sound.
6. Pressure force.
7. Non-pressure force.
8. Predicted positions and pressure.
9. Predicted velocities and density.

Each of these physical quantities are stored in a consecutive *float4* array. This array is kept sorted according to the position of the particles in the domain by employing a reference sort based on the particle positions data array. This data organization has the major advantage of providing efficient memory access, since corresponding physical quantities of particles are located in contiguous memory arrays. Since each of these data arrays is kept sorted according to the particle position, it is ensured that the physical quantities of particles which are close in space, are also allocated closely in the memory. This enables a more efficient caching strategy as presented in this work. Since, instead of caching the memory of the total memory required for each particle quantity, only the corresponding memory of the current physical quantity needs to be cached.

Moreover, this memory arrangement allows to significantly reduce the amount of data to be transferred between subdomains. In fact, in the PCISPH algorithm as presented in Algorithm 12, it is shown that typically each computation (such as *Compute Pressure Force* in line 19 of Algorithm 12) typically just changes one physical quantity. Due to the SOA particle representation, the data for pressure force is separated from the remaining quantities and hence, only the memory for pressure force needs to be exchanged between the respective subdomains. By that, the amount of memory to be transferred after *Compute Pressure Force* is reduced by factor nine, since the remaining eight data arrays remain unchanged.

Figure 10.3 illustrates the proposed particle data representation. The squared indicate the halo region, the crosses donate the padding region. By employing this structure, only individual data arrays need be exchanged and the memory of unmodified physical quantities can be neglected. Since all the arrays are sorted according to the particle position in the domain, the memory chunk of the halo region can be directly copied to the neighboring padding region without any overhead like additional sorting after every data exchange.

Keeping particles sorted according to the position within this data structure is essential for this approach. For that purpose, a radix sort is applied after every advection step, i.e. after the particle positions are modified. Hence, particles need to be sorted once at the end of every time step. Table 10.1 provides separate timing measurements for each individual kernel function as described in Algorithm 12. Individual measurements are shown for 1 million up to 32 million particles per GPU. It can be assessed that with increasing number of particles, the computational time spent on sorting is negligible. For 1 million particles, approximately 15% of

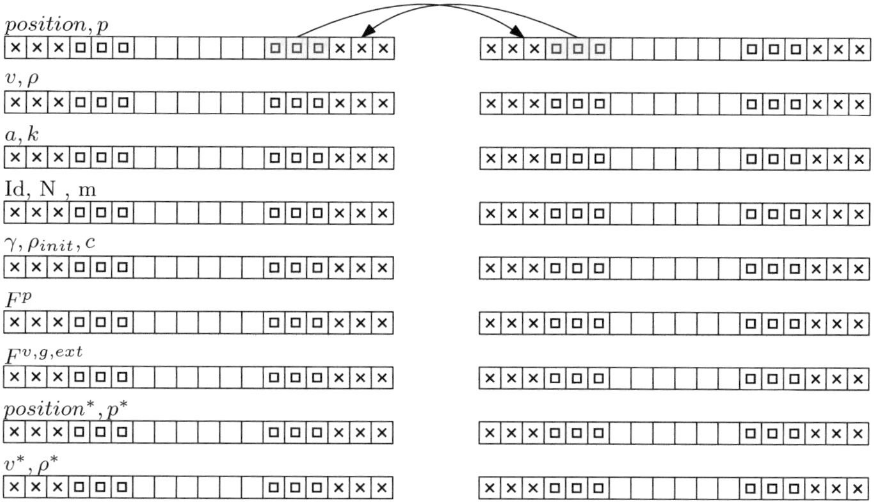

position, $p$

$v, \rho$

$a, k$

Id, N , m

$\gamma, \rho_{init}, c$

$F^p$

$F^{v,g,ext}$

position*, $p^*$

$v^*, \rho^*$

**Fig. 10.3** Structure of arrays (SOA) particle representation, squares and crosses donate halo and paddings regions, respectively. Only the required data array needs to be exchanged between the subdomains

**Table 10.1** Timing measurements of different kernel functions for a range of particles per GPU denoted in %

| # particles [mio] | 1 | 2 | 4 | 8 | 16 | 32 |
|---|---|---|---|---|---|---|
| Boundary | 6 | 3.5 | 2.1 | 1.1 | 0.8 | 0.6 |
| Nonpressure | 8.4 | 10.5 | 11.9 | 12.3 | 12.8 | 13 |
| Predict density | 36.7 | 43.7 | 45.5 | 47.3 | 47.8 | 48.2 |
| Pressure | 2.5 | 1.4 | 1 | 0.6 | 0.3 | 0.2 |
| Pressure correction | 26.4 | 30.5 | 33.8 | 35.7 | 36.5 | 36.8 |
| Pressure force | 2.4 | 1.3 | 0.8 | 0.4 | 0.3 | 0.3 |
| Advection | 2.2 | 1.2 | 0.8 | 0.4 | 0.3 | 0.2 |
| Sort | 15.4 | 7.9 | 4.1 | 2.2 | 1.2 | 0.7 |

the computational time for an individual time step is dedicated to sorting, whereas only 2% is required for a particle number of 8 million. Starting with 16 million particles, the time spent on sorting is less than 1% as compared to the remaining kernel functions.

## 10.3.4 Optimizing Exchange of Halos

The process of exchanging halos constitutes the major bottleneck of any parallelization method for SPH. Especially the inner iterations of PCISPH yield an increased amount of synchronization points. Since in every iteration the particle values are frequently updated, the halos need to be exchanged between GPUs after every particle modification. In total, this yields $2 + K \cdot 4$ halo exchange processes, where $K$ refers to the number of iterations until it converges (typically 3–5 iterations).

In order to accelerate this bottleneck, the dedicated computations of PCISPH are classified into two categories: (1) computations that only change discrete particle values (such as density or pressure), and, (2) computations that alter the position of particles. Based on these classification, two distinct methods for exchanging halos are presented.

The first method is applied when only particle values are modified. Here, corresponding offsets for the halo and padding regions of each subdomain are computed. If particle positions are not modified, the number of particles within every subdomain remains unchanged. Therefore, it is not required to allocate new memory which would cause additional overhead. Hence, the memory of the halo region of each subdomain can be dispatched to the corresponding neighboring padding region. Since the padding and halo regions are independent of each other, all memory transfers can be performed asynchronously among all subdomains. The time spent on exchanging halos after computations that do not alter positions of particles is therefore not increasing with a larger number of subdomains.

For computations that alter positions of particles this method is not applicable, since the number of particles located inside the halo and padding region may be modified and hence, the amount of memory to be transferred needs to be recomputed. For that purpose, the following algorithm is applied:

1. The data inside the padding region, which will be overwritten by the halo data from the neighboring subdomain, is deleted.
2. After particle positions are altered, it may lead to a new number of particles in the halo regions. This leads to a new number of particles located inside the neighboring paddings regions which overall results in a new number of particles inside the subdomain. For that purpose, new memory for the padding regions based on the neighboring halos needs to be allocated.
3. Memory of halos and paddings from right to left subdomains is exchanged (i.e. memory of left paddings is dispatched to memory of neighboring subdomains right halo).
4. The exchange of memory between left and right halos is conducted. Since the additional memory is allocated at the end of each data array, the free memory chunk is always located at the end of each array. Therefore, the right halos of neighboring subdomains are dispatched to the very end of each array instead of the beginning of the array.
5. After these memory exchanges, the particles are sorted according to their positions in the domain. Halo and padding regions are updated based on the new number of particles.

These steps are also illustrated in Fig. 10.4.

Note that this approach suggests that there is a need for frequent allocation and deallocation of memory due to the inherent moving nature of particles. To reduce this performance loss caused by this memory handling, a dedicated particle

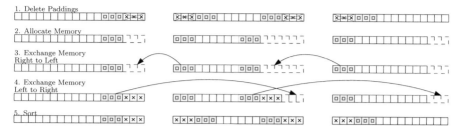

**Fig. 10.4** Exchange of halos after computations that alter the position of particles. Paddings are denoted by crosses, halos by squares

memory data structure is implemented, which avoids these frequent allocations and deallocations of memory. Here, in case particles need to be deleted, the memory is not actually deallocated, instead just the internal *size*, i.e. the number of particles currently in the subdomain, is updated. Hence, the actual *capacity*, i.e. the allocated memory, remains unchanged. Once new particles are added to the subdomain, no new memory needs to be allocated, but only the internal *size* is updated again. Actual memory is only deallocated once the difference between *size* and *capacity* is larger than a certain threshold (which is chosen as 20% in the proposed implementation). Hence, a trade off between memory consumption and performance is applied.

Overall, this results in significantly improved scalability of the PCISPH method. Evaluations summarized in the next section confirm this improvement.

## 10.4   Experimental Evaluations

In order to evaluate the performance of the proposed scheme, the described methods have been implemented on top of an industrial PCISPH solver. Emphasis is on measuring the parallel performance, i.e. the scalability on multi-GPU systems. Correspondingly obtained experimental results are summarized in this section. In the following, the employed experimental setup is described in detail.

### 10.4.1   Experimental Setup

The performance has been evaluated by means of speedup and efficiency using strong scaling and weak scaling for a different number of particles on different number of GPUs. Strong scaling ($S_s$), weak scaling ($S_w$), and efficiency ($E$) are defined as

$$S_s = \frac{T_s(n)}{T_p(n, p)},$$
(10.3)

$$S_w = \frac{T_s(n)\,p}{T_p(n,\,p)}, \text{ and} \qquad\qquad (10.4)$$

$$E = \frac{S_s}{p}, \qquad\qquad (10.5)$$

respectively, where $n$ is the size of the input (number of particles in the system), $T_S$ the execution time of the single GPU implementation, $T_p$ the execution time of the multi-GPU implementation, and $p$ the number of used GPUs. Strong scaling determines the speedup for a fixed problem size, whereas weak scaling provides measures to determine the runtime for a fixed number of particles per GPU.

As test cases, two scenarios have been considered: (1) Kleefman's dam break scenario by [45], which is a well-established test case, and, (2) a scenario from the automotive industry, in which water splashing introduced by a car driving through water puddles is simulated. All test cases have been simulated on nodes composed of up to ten GPUs of type Nvidia GTX 1080 Ti, which contain 3584 CUDA cores with a memory bandwidth of 484 GB/s. The source code was compiled on CentOs 6 using gcc v4.5.3 and the CUDA Toolkit v9.1.

The results obtained for the dam break scenario are summarized next.

### *10.4.2 Dam Break Simulation*

The dam break test case (as shown in Fig. 10.5) is a well-established test scenario for CFD methods. It constitutes a major challenge for particle-based methods, since it is an inherently imbalanced problem. In the beginning of the simulation, all particles are clustered on one side, while by the end of the simulation they are equally distributed among the domain.

**Fig. 10.5** Dam break test case with velocity profile

**Table 10.2** Speedup and efficiency obtained for the dam break test case

| # particles | 2 GPUs | 4 GPUs | 8 GPUs | 16 GPUs |
|---|---|---|---|---|
| *(a) Speedup* | | | | |
| 12 mio | 1.79 | 3.12 | 4.32 | 3.84 |
| 28 mio | 1.89 | 3.34 | 4.81 | 4.74 |
| 48 mio | 1.92 | 3.61 | 5.15 | 6.15 |
| *(b) Efficiency* | | | | |
| 12 mio | 0.89 | 0.78 | 0.54 | 0.24 |
| 28 mio | 0.95 | 0.83 | 0.60 | 0.30 |
| 48 mio | 0.96 | 0.90 | 0.64 | 0.38 |

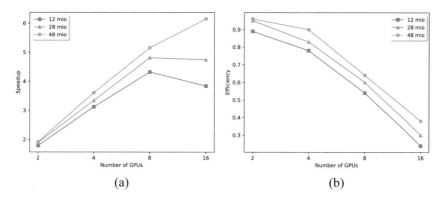

**Fig. 10.6** Speedup and efficiency obtained for the dam break test case. (**a**) Speedup. (**b**) Efficiency

The obtained experimental results in terms of speedup and efficiency are summarized in Table 10.2 as well as visualized in Fig. 10.6.

It is shown that higher speedup is achieved when a larger number of particles is employed. Moreover, the efficiency drops with higher numbers of GPUs. This essentially results from the highly unbalanced nature of the dam break scenario, which necessitates to frequently conduct a domain decomposition correction (as described in Sect. 10.3.1). The relative overhead introduced by this technique is larger with a smaller number of particles per GPU. However, the obtained results show that the inherent synchronization burden can be significantly reduced by the presented methods. It is shown that PCISPH can get accelerated by the multi-GPU technique even for relatively small number of particles, e.g. for a 12 mio. particle simulation a speedup of 3.12 is achieved on four GPUs. For a larger number of particles, a close to optimal efficiency of 0.9 can be achieved on four GPUs.

The results for the weak scaling approach, i.e. a fixed number of particles per GPU, are visualized in Fig. 10.7. Here, a problem size ranging from 1 mio. up to 32 mio. particles are considered. It can be assessed that, with smaller numbers of particles, the speedup drops significantly. However, with an increasing number of particles, satisfying results close to the ideal curve are achieved. More precisely, with 16 mio. and 32 mio. particles, a speedup of 14.8 and 15.6, respectively, is

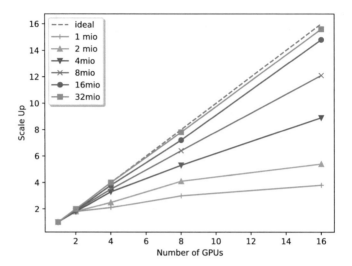

**Fig. 10.7** Speedup obtained for the dam break test case using weak scaling

obtained. These results also indicate that this approach will scale well for a larger number of GPUs if the number of particles per GPU is chosen close to the ideal number.

### 10.4.3   Water Splashing Simulation

The water splashing case is a scenario from automotive industry, in which splashing water induced by wet roads is simulated. This splashing water can enter the car and lead to damage, e.g. rusting on metal object or water entering the engine. Moreover, splashing pressure applied to different parts cause potential risks of object deformation.

The presented simulation technique is applied to this industrial scenario to simulate the splashing of the water near the tires and through the chassis into the car body. The car geometry has dimensions of approximately $x = 4\,\text{m}$, $y = 2\,\text{m}$, $z = 1.2\,\text{m}$, and is moving along a straight line in the X-direction at a constant speed of 30 km/h. In this scenario, the puddle consists of water with a density $\rho = 1000\,\text{kg/m}^3$ and a kinematic viscosity $\nu = 8.94 \times 10^{-4}\,\text{m}^2/\text{s}$. At the beginning of the simulation, the car is placed outside of the water puddle such that the front tires are slightly in contact with the edge of the puddle. The puddle is a rectangular region of 6 m length and 80 mm height. The sides of the domain are defined as open boundaries, i.e. water particles that are splashed to the sides of the domain will be removed from the simulation (such open boundaries are also referred to as *Sinks*). Simulations are executed such that the car body moves to the end of the axial domain.

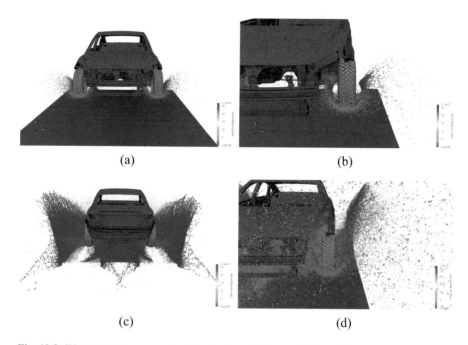

**Fig. 10.8** Water splashing scenario with velocity profile in two different time steps. (**a**) Front view at $t = 0.1$ s. (**b**) Tire view at $t = 0.1$ s. (**c**) Back view at $t = 0.5$ s. (**d**) Tire view at $t = 0.5$ s

**Table 10.3** Speedup and efficiency obtained for the water splashing scenario

| # particles | 2 GPUs | 4 GPUs | 8 GPUs | 16 GPUs |
|---|---|---|---|---|
| *(a) Speedup* | | | | |
| 12 mio | 1.60 | 3.11 | 5.60 | 2.87 |
| 24 mio | 1.63 | 3.20 | 5.75 | 5.06 |
| *(b) Efficiency* | | | | |
| 12 mio | 0.80 | 0.78 | 0.70 | 0.18 |
| 24 mio | 0.82 | 0.80 | 0.72 | 0.32 |

In Fig. 10.8, the simulation results with a velocity profile are shown at time step $t = 0.1$ s and $t = 0.5$ s. The scenario has been simulated using two different resolutions, employing 12 mio. and 24 mio. particles. The obtained experimental results for speedup and efficiency are summarized in Table 10.3 as well as visualized in Fig. 10.9. It is shown that there is no significant drop of efficiency up to eight GPUs. As compared to the dam break test case, in this scenario is evenly balanced, i.e. the particles are evenly distributed among to domain. Hence, the domain decomposition correction is not frequently applied, resulting in less overhead.

However, the speedup obtained for two GPUs is lower as compared to the dam break scenario. This essentially results from the open boundary condition, where particles need to be frequently deleted from the simulation when they enter the sides of the domain. This deletion of memory causes an additional synchronization step

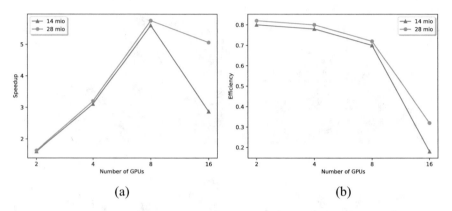

**Fig. 10.9** Speedup and efficiency obtained for the water splashing scenario. (**a**) Speedup. (**b**) Efficiency

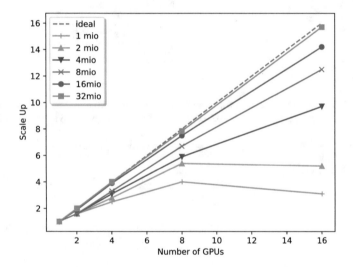

**Fig. 10.10** Speedup obtained for the water splashing scenario using weak scaling

over all devices, leading to slightly reduced speedup. However, this synchronization is conducted in parallel over all devices, hence the loss of speedup caused by this step is not increasing with larger number of GPUs.

The results obtained for the weak scaling approach are visualized in Fig. 10.10. A fixed number of particles per GPU is used in a range of 1 mio. to 32 mio. Similar to the dam break test case, it can be assessed that, with a larger number of particles, the obtained results are close to the ideal curve. With 16 mio. particles per GPU, a speedup of 14.2 is achieved on 16 GPUs. Employing 32 mio. particles, the obtained speedup is almost ideal with a value of 15.7. These results again indicate that the approach scales well also for a larger number of GPUs if these ideal number of particles per GPU are employed.

Overall, it is shown that the simulation time for industrial use cases can be significantly reduced by employing the presented accelerated PCISPH methods. Since PCISPH sequentially clearly outperforms other famous SPH variants such as WCSPH, the results show that by employing the proposed methods, PCISPH becomes suitable to simulate industrial use cases which involve a larger number of particles.

## 10.5  Summary

In this chapter, a distributed multi-GPU scheme for the PCISPH method is presented. The proposed scheme has been implemented on top of an industrial solver, including various optimization methodologies. To that end, a dedicated load balancing technique and approaches to overcome the severe synchronization problems of PCISPH are presented. The PCISPH method sequentially outperforms other SPH variants such as WCSPH by up to one order of magnitude. It is shown that the proposed methods allow to employ a distributed multi-GPU scheme to further accelerate the PCISPH method. Especially with a larger number of particles per GPU, speedups close to the ideal values could be achieved. Experimental evaluations on a dam break test case and an industrial water splash scenario confirm that the presented methods allow for PCISPH to be applied to real world scenarios involving a large number of particles.

As a future work, different methodologies for domain decomposition, such as presented by [43] will be considered. Here, different decomposition methods for a WCSPH solver are presented, in which, in the first two approaches, the domain is decomposed exclusively along x-axis and z-axis, respectively. In a third approach, the domain is decomposed along both axises. It is shown that, for this particular water entry scenario, the domain decomposition along both axises yields the least communication overhead. In [90], space filling curves are used for the purpose of domain decomposition and further extended for dynamic load balancing. For the conducted test cases it is shown that, especially for a large number of GPUs, performance scalability improves. These methods will be evaluated by means of the presented PCISPH solver to further assess the feasibility and applicability to large scale industrial scenarios. Moreover, multi resolution aspects and a more efficient methodology for *Sinks* (deletion of particles at runtime) and *Sources* (creation of particles at runtime) will be investigated.

# Part V
# Conclusion

# Chapter 11
# Conclusion

This book substantially contributed to the goal of making dedicated *Computational Fluid Dynamics* (CFD) approaches more efficient and applicable to real world problems—trying to avoid scenarios where fluid flow problems can only be modeled by employing simplifications. In fact, investigating the dedicated CFD approaches of *grid-based methods*, *volumetric decomposition methods*, and *particle-based methods* from a *High Performance Computing* (HPC) perspective led to novel methodologies that significantly outperform the current state of the art. These accomplishments are summarized as follows:

A dedicated overset grid scheme has been proposed for the *grid-based Finite Difference Method* (FDM). The proposed scheme employs different levels of discretization, allowing to model the area of interest in high resolution, whereas areas of less interest are discretized in lower resolution. Following this scheme eventually allows to significantly reduce the number of required disretization points by still maintaining the accuracy of the simulation. Moreover, this scheme ensures that even tiny layers of material are discretized by at least two layers of grid points, which is an essential aspect in a range of simulations. The proposed scheme led to a novel simulation tool, which allows to accurately simulate a dedicated application of the automotive industry, in which corrosion protection is applied to car bodies.

For *volumetric decomposition methods*, significant improvements in terms of computational time have been achieved. To that end, a framework has been developed, which allows to compute decompositions in parallel, rather than computing them sequentially for each discrete time step of a simulation. The framework employs parallelism on a *threading* level (i.e. on shared memory architectures), as well as on a *process* level (i.e. on distributed memory architectures). Moreover, a *hybrid* parallel scheme is presented, which combines the two levels of parallelism—resulting in significant speedup. Further optimizations to workload distributions, memory management and load balancing are considered, which eventually allow to overcome the drawback of volumetric decomposition methods employed thus far.

K. Verma, R. Wille, *High Performance Simulation for Industrial Paint Shop Applications*, https://doi.org/10.1007/978-3-030-71625-7_11

Finally, this book has substantially contributed to dealing with the complexities of *particle-based methods*. To that end, methods for *General Purpose Computations on Graphics Processing Units* (GPGPU) are utilized that allow for massively parallel computations on GPUs. These methods are extended by a dedicated multi-GPU architecture, allowing to distribute the respective computations on multiple GPUs. For that purpose, spatial subdivisions are employed that partition the computational domain into individual subdomains. These subdomains are then individually executed in parallel on multiple GPUs. Dedicated load balancing and optimization techniques are presented, that allow for efficient hardware utilization and, by that, result into significant speedup.

The investigations and developments conducted within the context of this book led to a variety of methods that significantly outperform current state of the art. This is confirmed by both, academic use cases, as well as real world industrial scenarios. Besides these methods and improvements, parts of this book have also been published in international journals, as well as proceedings of international conferences.

For future work, various algorithmic aspects shall be investigated to further improve the scalability and applicability of the dedicated CFD approaches. In this regard, especially adaptive resolution aspects yield promising directions. It is shown that many real world applications yield areas with varying interest and potential of error. This suggests that adaptive resolution approaches may also allow to further reduce the complexity of e.g. particle-based CFD approaches.

# References

1. G. Amdahl, Validity of the single processor approach to achieving large scale computing capabilities, in *Proceedings of the April 18–20, 1967, Spring Joint Computer Conference* (Association for Computing Machinery, New York, 1967), pp. 483–485
2. J.D. Anderson, *Computational Fluid Dynamics*, vol. 206 (Springer, New York, 1995)
3. T.J. Baker, Mesh adaptation strategies for problems in fluid dynamics. Finite Elem. Anal. Des. **25**(3–4), 243–273 (1997)
4. M. Becker, M. Teschner, Weakly compressible SPH for free surface flows, in *Proceedings of the 2007 ACM SIGGRAPH/Eurographics Symposium on Computer Animation* (Eurographics Association, Genoa, 2007), pp. 209–217
5. T. Belytschko, Y. Guo, W.K. Liu, S.P. Xiao, A unified stability analysis of meshless particle methods. Int. J. Numer. Methods Eng. **48**(9), 1359–1400 (2000)
6. L. Besra, M. Liu, A review on fundamentals and applications of electrophoretic deposition (EPD). Progr. Mater. Sci. **52**(1), 1–61 (2007)
7. M.M.A. Bhutta, N. Hayat, M.H. Bashir, A.R. Khan, K.N. Ahmad, S. Khan, CFD applications in various heat exchangers design: a review. Appl. Therm. Eng. **32**, 1–12 (2012)
8. A.R. Boccaccini, I. Zhitomirsky, Application of electrophoretic and electrolytic deposition techniques in ceramics processing. Curr. Opin. Solid State Mater. Sci. **6**(3), 251–260 (2002)
9. S. Bysko, J. Krystek, S. Bysko, Automotive paint shop 4.0. Comput. Ind. Eng. **139**, 105546 (2020)
10. A. Cheng, D. Cheng, Heritage and early history of the boundary element method. Eng. Anal. Bound. Elem. **29**, 268–302 (2005)
11. Y. Chikazawa, S. Koshizuka, Y. Oka, A particle method for elastic and visco-plastic structures and fluid-structure interactions. Comput. Mech. **27**(2), 97–106 (2001)
12. A. Chorin, Numerical solution of the navier-stokes equations. Math. Comput. **22**, 10 (1968)
13. A.J. Chorin, Numerical solution of the Navier-Stokes equations. Math. Comput. **22**(104), 745–762 (1968)
14. A. Chorin, On the convergence of discrete approximations to the Navier-Stokes equations. Math. Comput. **23**, 341–341 (1969)
15. L. Clarke, I. Glendinning, R. Hempel, The MPI message passing interface standard, in *Programming Environments for Massively Parallel Distributed Systems* (Springer, New York, 1994), pp. 213–218
16. P. Constantin, C. Foias, *Navier-Stokes Equations* (University of Chicago Press, Chicago, 1988)
17. S.J. Cummins, M. Rudman, An SPH projection method. J. Computat. Phys. **152**(2), 584–607 (1999)

18. L. Dagum, R. Menon, OpenMP: an industry standard API for shared-memory programming. IEEE Comput. Sci. Eng. **5**(1), 46–55 (1998)
19. A. Datta, J. Sitaraman, I. Chopra, J.D. Baeder, CFD/CSD prediction of rotor vibratory loads in high-speed flight. J. Aircraft **43**(6), 1698–1709 (2006)
20. J.M. Domínguez, A.J.C. Crespo, D. Valdez-Balderas, B.D. Rogers, M. Gómez-Gesteira, New multi-GPU implementation for smoothed particle hydrodynamics on heterogeneous clusters. Comput. Phys. Commun. **184**(8), 1848–1860 (2013)
21. H. Doraiswamy, V. Natarajanb, Efficient algorithms for computing Reeb graphs. Comput. Geom. **42**, 606–616 (2009)
22. L. Dumas, CFD-based optimization for automotive aerodynamics, in *Optimization and Computational Fluid Dynamics* (Springer, New York, 2008), pp. 191–215
23. R. Eymard, T. Gallouët, R. Herbin, Finite volume methods, in *Handbook of Numerical Analysis*, vol. 7 (North-Holland, Amsterdam, 2000), pp. 713–1018
24. A. Ferrari, M. Dumbser, E.F. Toro, A. Armanini, A new 3d parallel sph scheme for free surface flows. Comput. Fluids **38**(6), 1203–1217 (2009)
25. J. Fung, S. Mann, Using multiple graphics cards as a general purpose parallel computer: applications to computer vision, in *Proceedings of the 17th International Conference on Pattern Recognition, 2004. ICPR 2004*, vol. 1 (IEEE, New York, 2004), pp. 805–808
26. J. Fung, F. Tang, S. Mann, Mediated reality using computer graphics hardware for computer vision, in *Proceedings. Sixth International Symposium on Wearable Computers* (IEEE, New York, 2002), pp. 83–89
27. E. Gabriel, G.E. Fagg, G. Bosilca, T. Angskun, J.J. Dongarra, J.M. Squyres, V. Sahay, P. Kambadur, B. Barrett, A. Lumsdaine, R.H. Castain, D.J. Daniel, R.L. Graham, T.S. Woodall, Open MPI: goals, concept, and design of a next generation MPI implementation, in *11th European PVM/MPI Users' Group Meeting*, Budapest, Hungary, September 2004, pp. 97–104
28. R.A. Gingold, J.J. Monaghan, Smoothed particle hydrodynamics: theory and application to non-spherical stars. Mon. Not. R. Astron. Soc. **181**(3), 375–389 (1977)
29. M.T. Goodrich, Intersecting line segments in parallel with an output-sensitive number of processors. SIAM J. Comput. **20**(4), 737–755 (1991)
30. M.T. Goodrich, J.-J. Tsay, D.E. Vengroff, J.S. Vitter, External-memory computational geometry, in *IEEE 34th Annual Foundations of Computer Science* (1993), pp. 714–723
31. W. Gropp, W.D. Gropp, E. Lusk, A. Skjellum, Argonne distinguished fellow Emeritus Ewing Lusk, in *Using MPI: Portable Parallel Programming with the Message-Passing Interface*, vol. 1 (MIT Press, Cambridge, 1999)
32. C. Grossmann, H.-G. Roos, M. Stynes, *Numerical Treatment of Partial Differential Equations*, vol. 154 (Springer, New York, 2007)
33. K. Gustafson, Domain decomposition, operator trigonometry, Robin condition. Contemp. Math. **218**, 432–437 (1998)
34. T. Harada, S. Koshizuka, Y. Kawaguchi, Smoothed particle hydrodynamics on GPUs, in *Proceedings of the 5th International Conference Computer Graphics* (2007), pp. 63–70
35. S. Heo, S. Koshizuka, Y. Oka, Numerical analysis of boiling on high heat-flux and high subcooling condition using MPS-MAFL. Int. J. Heat Mass Transf. **45**(13), 2633–2642 (2002)
36. A. Hérault, G. Bilotta, R.A. Dalrymple, SPH on GPU with CUDA. J. Hydraul. Res. **48**(S1), 74–79 (2010)
37. D. Hietel, K. Steiner, J. Struckmeier, A finite-volume particle method for compressible flows. Math. Models Methods Appl. Sci. **10**(09), 1363–1382 (2000)
38. C. Hirsch, The finite difference method for structured grids, in *Numerical Computation of Internal and External Flows*, ed. by C. Hirsch, Chapter 4, 2nd edn. (Butterworth-Heinemann, Oxford, 2007), pp. 145–201
39. M. Ihmsen, N. Akinci, M. Becker, M. Teschner, A parallel SPH implementation on multi-core CPUS. Comput. Graph. Forum **30**(1), 99–112 (2011)
40. M.R.I. Islam, C. Peng, A total Lagrangian SPH method for modelling damage and failure in solids. Int. J. Mech. Sci. **157–158**, 498–511 (2019)

41. R. Issa, E.S. Lee, D. Violeau, D. Laurence, Incompressible separated flows simulations with the smoothed particle hydrodynamics gridless method. Int. J. Numer. Methods Fluids **47**(10–11), 1101–1106 (2005)
42. S. Jamshed, *Using HPC for Computational Fluid Dynamics: A Guide to High Performance Computing for CFD Engineers.* (Academic, Cambridge, 2015)
43. Z. Ji, F. Xu, A. Takahashi, Y. Sun, Large scale water entry simulation with smoothed particle hydrodynamics on single- and multi-GPU systems. Comput. Phys. Commun. **209**, 1–12 (2016)
44. F. Ju, J. Li, G. Xiao, J. Arinez, Modeling quality propagation in automotive paint shops: an application study. IFAC Proc. **46**(9), 1890–1895 (2013)
45. K.M.T. Kleefsman, G. Fekken, A.E.P. Veldman, B. Iwanowski, B. Buchner, A volume-of-fluid based simulation method for wave impact problems. J. Comput. Phys. **206**(1), 363–393 (2005)
46. C. Kleinstreuer, *Modern Fluid Dynamics* (Springer, New York, 2018)
47. A. Kolb, N. Cuntz, Dynamic particle coupling for GPU-based fluid simulation, in *Proceedings of the Symposium on Simulation Technique.* Citeseer (2005), pp. 722–727
48. S. Koshizuka, Y. Oka, Moving-particle semi-implicit method for fragmentation of incompressible fluid. Nucl. Sci. Eng. **123**(3), 421–434 (1996)
49. S. Koshizuka, H. Ikeda, Y. Oka, Numerical analysis of fragmentation mechanisms in vapor explosions. Nucl. Eng. Des. **189**(1–3), 423–433 (1999)
50. H.-P. Kriegel, T. Brinkhoff, R. Schneider, The combination of spatial access methods and computational geometry in geographic database systems, in *Symposium on Spatial Databases* (Springer, New York, 1991), pp. 5–21
51. E.-S. Lee, C. Moulinec, R. Xu, D. Violeau, D. Laurence, P. Stansby, Comparisons of weakly compressible and truly incompressible algorithms for the sph mesh free particle method. J. Comput. Phys. **227**(18), 8417–8436 (2008)
52. E.-S. Lee, D. Violeau, R. Issa, S. Ploix, Application of weakly compressible and truly compressible SPH to 3-D water collapse in waterworks. J. Hydraul. Res. **2010**, 50–60 (2010)
53. M.B. Liu, G.R. Liu, Smoothed particle hydrodynamics (SPH): an overview and recent developments. Arch. Comput. Methods Eng. **17**(1), 25–76 (2010)
54. M. McKenney, T. McGuire, A parallel plane sweep algorithm for multi-core systems, in *International Conference on Advances in Geographic Information Systems* (ACM, New York, 2009), pp. 392–395
55. M. Menon, S. Baig, K. Verma, Computational analysis of spray pre-treatment in automotive applications. SAE Technical Paper. SAE International, 04 2020
56. J. Milnor, *Morse Theory* (Princeton University Press, Princeton, 1963), p. 51
57. J.J. Monaghan, Simulating free surface flows with SPH. J. Comput. Phys. **110**(2), 399–406 (1994)
58. J.J. Monaghan, Smoothed particle hydrodynamics. Rep. Prog. Phys **68**, 1703–1759 (2005)
59. M. Müller, D. Charypar, M. Gross, Particle-based fluid simulation for interactive applications, in *Symposium on Computer Animation*, vol. 2003 (2003), pp. 154–159
60. B.R. Munson, T.H. Okiishi, W.W. Huebsch, A.P. Rothmayer, *Fluid Mechanics* (Wiley, Singapore, 2013)
61. F.N. Jones, M.E. Nichols, S.P. Pappas, Electrodeposition coatings, in *Organic Coatings: Science and Technology* (Wiley, Chichester, 2017), pp. 374–384
62. T.Y. Na, *Computational Methods in Engineering Boundary Value Problems* (Academic, New York, 1980)
63. H. Nagai, Y. Onishi, K. Amaya, Exact paint resistance and deposition models for accurate numerical electrodeposition simulation. Trans. Jpn. Soc. Mech. Eng. Ser. A **78**(794), 1446–1461 (2012)
64. J.R. Nagel, Numerical solutions to poisson equations using the finite-difference method [education column]. IEEE Antenn. Propag. Mag. **56**(4), 209–224 (2014)
65. J. Nickolls, I. Buck, M. Garland, K. Skadron, Scalable parallel programming with CUDA. Queue **6**(2), 40–53 (2008)

66. J. Nievergelt, F.P. Preparata, Plane-sweep algorithms for intersecting geometric figures. Commun. ACM **25**, 739-747 (1982)
67. T. Norton, D.-W. Sun, J. Grant, R. Fallon, V. Dodd, Applications of computational fluid dynamics (CFD) in the modelling and design of ventilation systems in the agricultural industry: a review. Bioresour. Technol. **98**(12), 2386–2414 (2007)
68. Nvidia Corp., Nvidia CUDA C Programming Guide. Technical Report 4.2, Nvidia Corp. Org., Santa Clara, CA, 2012
69. A. Oberman, I. Zwiers, Adaptive finite difference methods for nonlinear elliptic and parabolic partial differential equations with free boundaries. J. Sci. Comput. **68**, 12 (2014)
70. A. Osseyran, M. Giles, *Industrial Applications of High-Performance Computing: Best Global Practices* (Chapman & HallCRC, Boca Raton, 2015)
71. C. Peng, C. Bauinger, K. Szewc, W. Wu, H. Cao, An improved predictive-corrective incompressible smoothed particle hydrodynamics method for fluid flow modelling. J. Hydrodyn. **31**(4), 654–668 (2019)
72. H. Peters, O. Schulz-Hildebrandt, N. Luttenberger, Fast in-place sorting with cuda based on bitonic sort, in *International Conference on Parallel Processing and Applied Mathematics* (Springer, New York, 2009), pp. 403–410
73. M.J. Quinn, Parallel programming in C with MPI and OpenMP (2003)
74. P.W. Randles, L.D. Libersky, Smoothed particle hydrodynamics: some recent improvements and applications. Comput. Methods Appl. Mech. Eng. **139**(1–4), 375–408 (1996)
75. E. Rustico, G. Bilotta, A. Herault, C.D. Negro, G. Gallo, Advances in multi-GPU smoothed particle hydrodynamics simulations. IEEE Trans. Parall. Distrib. Syst. **25**(1), 43–52 (2012)
76. I.F. Sbalzarini, J.H. Walther, M. Bergdorf, S.E. Hieber, E.M. Kotsalis, P. Koumoutsakos, PPM - a highly efficient parallel particle-mesh library for the simulation of continuum systems. J. Comput. Phys. **215**(2), 566–588 (2006)
77. T.C. Schroeder, Peer-to-peer and unified virtual addressing, in *GPU Technology Conference, NVIDIA* (2011)
78. S. Shadloo, G. Oger, D.L. Touzé, Smoothed particle hydrodynamics method for fluid flows, towards industrial applications: motivations, current state, and challenges. Comput. Fluids **136**, 11–34 (2016)
79. S. Shao, E.Y.M. Lo, Incompressible sph method for simulating Newtonian and non-Newtonian flows with a free surface. Adv. Water Resour. **26**(7), 787–800 (2003)
80. A. Shimura, Y. Onishi, K. Amaya, Electrodeposition simulation considering history dependency, in *The Proceedings of the Computational Mechanics Conference* (2016)
81. B. Solenthaler, R. Pajarola, Predictive-corrective incompressible SPH, in *ACM SIGGRAPH 2009 Papers* (2009), pp. 1–6
82. G. Strang, G.J. Fix, *An analysis of the Finite Element Method*, vol. 212 (Prentice-Hall, Englewood Cliffs, NJ, 1973)
83. B. Strodthoff, M. Schifko, B. Juettler, Horizontal decomposition of triangulated solids for the simulation of dip-coating processes. Comput.-Aided Des. **43**, 1891–1901 (2011)
84. J.W. Swegle, D.L. Hicks, S.W. Attaway, Smoothed particle hydrodynamics stability analysis. J. Comput. Phys. **116**(1), 123–134 (1995)
85. A. Taflove, S.C. Hagness, *Computational Electrodynamics: The Finite-Difference Time-Domain Method* (Artech House, Boston, 2005)
86. R. Temam, Navier-Stokes Equations: Theory and Numerical Analysis, vol. 343 (American Mathematical Society, Providence, 2001)
87. J.F. Thompson, Grid generation techniques in computational fluid dynamics. AIAA J. **22**(11), 1505–1523 (1984)
88. Y. Tominaga, A. Mochida, R. Yoshie, H. Kataoka, T. Nozu, M. Yoshikawa, T. Shirasawa, AIJ guidelines for practical applications of CFD to pedestrian wind environment around buildings. J. Wind Eng. Ind. Aerodyn. **96**(10–11), 1749–1761 (2008)
89. J. Glaser, T.D. Nguyen, J.A. Anderson, P. Lui, F. Spiga, J.A. Millan, D.C. Morse, S.C. Glotzer, Strong scaling of general-purpose molecular dynamics simulations on GPUs. Comput. Phys. Commun. Elsevier. **192**, 97–107 (2015)

90. S. Tsuzuki, T. Aoki, Effective dynamic load balance using space-filling curves for large-scale SPH simulations on gpu-rich supercomputers, in *2016 7th Workshop on Latest Advances in Scalable Algorithms for Large-Scale Systems (ScalA)* (IEEE, New York, 2016), pp. 1–8
91. O. Van der Biest, L. Vandeperre, Electrophoretic deposition of materials. Ann. Rev. Mater. Sci. **29**, 327–352 (2003)
92. K. Verma, K. Szewc, R. Wille, Advanced load balancing for SPH simulations on multi-GPU architectures, in *IEEE High Performance Extreme Computing Conference* (2017), pp. 1–7
93. K. Verma, C. Peng, K. Szewc, R. Wille, A Multi-GPU PCISPH Implementation with efficient memory transfers, in *IEEE High Performance Extreme Computing Conference* (2018), pp. 1–7
94. K. Verma, L. Ayuso, R. Wille, Parallel simulation of electrophoretic deposition for industrial automotive applications, in *International Conference on High Performance Computing & Simulation* (2018), pp. 1–8
95. K. Verma, J. Oder, R. Wille, Simulating industrial electrophoretic deposition on distributed memory architectures, in *Euromicro Conference on Parallel, Distributed, and Network-Based Processing* (2019), pp. 1–8
96. K. Verma, H. Cao, P. Mandapalli, R. Wille, Modeling and simulation of electrophoretic deposition coatings. J. Comput. Sci. **41**, 101075 (2020)
97. K. Verma, C. McCabe, C. Peng, R. Wille, A PCISPH implementation using distributed multi-GPU acceleration for simulating industrial engineering applications. Int. J. High Perform. Comput. Appl. **34**(4), 450–464 (2020)
98. H.K. Versteeg, W. Malalasekera, *An Introduction to Computational Fluid Dynamics: The Finite Volume Method* (Pearson Education, New York, 2007)
99. R. Vignjevic, J. Reveles, J. Campbell, SPH in a total lagrangian formalism. Comput. Model. Eng. Sci. **14**, 181–198 (2006)
100. Q. Wu, C. Yang, T. Tang, K. Lu, Fast parallel cutoff pair interactions for molecular dynamics on heterogeneous systems. Tsinghua Sci. Technol. **17**(3), 265–277 (2012)
101. B. Xia, D.-W. Sun, Applications of computational fluid dynamics (CFD) in the food industry: a review. Comput. Electron. Agric. **34**(1–3), 5–24 (2002)
102. Z. Xu, G. Anyasodor, Y. Qin, Painting of aluminium panels - state of the art and development issues. MATEC Web of Conf. **21**, 05012 (2015)
103. H.Y. Yoon, S. Koshizuka, Y. Oka, Direct calculation of bubble growth, departure, and rise in nucleate pool boiling. Int. J. Multiphase Flow **27**(2), 277–298 (2001)
104. P. Zaspel, M. Griebel, Massively parallel fluid simulations on Amazon's HPC Cloud, in *First International Symposium on Network Cloud Computing and Applications* (2011), pp. 73–78
105. J.-M. Zhang, L. Zhong, B. Su, M. Wan, J.S. Yap, J.P.L. Tham, L.P. Chua, D.N. Ghista, R.S. Tan, Perspective on CFD studies of coronary artery disease lesions and hemodynamics: a review. Int. J. Numer. Methods Biomed. Eng. **30**(6), 659–680 (2014)

# Index

© The Author(s), under exclusive license to Springer Nature Switzerland AG 2021
K. Verma, R. Wille, *High Performance Simulation for Industrial Paint Shop Applications*, https://doi.org/10.1007/978-3-030-71625-7

Printed in the United States
by Baker & Taylor Publisher Services